TROPICAL GRAZING LANDS:

Communities and Constituent Species

TROPICAL GRAZING LANDS:

Communities and Constituent Species

Robert Orr Whyte

1974

Dr. W. JUNK b.v. - PUBLISHERS - THE HAGUE

ISBN 90 6193 020 0

Cover design: Max Velthuys
Printed in The Netherlands by Drukkerij Dijkstra Niemeyer b.v., Groningen

In memory of

Sir George and Lady Stapledon

and for the

Delegates to the

Fourth International Grassland Congress

Aberystwyth, 1937

when grassland research became international

CONTENTS

CONTENTS OF CHAPTERS

1 INTRODUCTION

1.1 Objective and scope

These chapters are not intended as a treatise on range management of tropical grazing lands, nor on the agronomy of sown pastures and cultivated fodder crops, aspects of applied science that are only one stage above scientific farming and use of land. An attempt is made to present the vegetation sciences, or botany and ecology in their widest aspects, as the essential background for their application in economic land use and plant and animal husbandry. The basic thesis is that science is global, but that its practical application is restricted to specific biological and socio-economic habitats.

The region covered is termed the intertropical zone. This comprises the equatorial latitudes, the tropics and subtropics. It is not only that part of the globe lying between latitudes 30° north and south of the Equator, as proposed by some writers (DAVIES, W., 1960; repeated by DAVIES & SKIDMORE, 1966). Rather is it those regions of the world in which, at the lower elevations, certain families and members of the Gramineae and Leguminosae grow wild or can be cultivated. It may be a matter for discussion whether these plants in their wild communities or sown or planted crop mixtures are better indicators of a biological environment than the instruments and criteria of the meteorologists.

Subtropical and even tropical species extend down to South Africa, for possible palaeobotanical reasons discussed below (3.1.1.). Corrientes in Argentina is included, but Rio Grande do Sul in Brazil is not (see 3.2.4). The brigalow lands in Australia (COALDRAKE, 1970) are regarded there as subtropical, and may thus be included; it is, however, found that in north-eastern Australia, the differences between the tropical and subtropical zones necessitate quite different approaches to land management. Subtropical areas with a cold season and frosts, the annual alternation on the same site of tropical and temperate conditions, are among the most difficult environments in which to develop sown pastures (Taiwan, Korea, Japan). Results of work in these areas cannot be extrapolated to fully subtropical conditions, and vice versa.

The botanical sciences as they apply to the tropical grazing lands are considered primarily in relation to that geographical entity, highly diverse and variable, that comprises the biologically intertropical belt lying along both sides of the Equator, but also in relation to the separate parts of that zone, namely:

a. Australia and the Pacific Islands
b. Asia, monsoonal and equatorial (insular)
c. Africa south of the Sahara, and
d. America, South, Central and the Caribbean

There are, of course, extensive areas of high mountain country to which the criteria for the tropical grazing land environments do not apply, and which represent a separate branch of vegetation ecology.

1.2 **Management of tropical grazing lands**

Most of the current research and published literature on the tropical grazing lands, natural and artificial, is concerned with what may be described as agronomic and managerial questions, with advice to the grazier or the farmer as the immediate objective. Thus it deals with management of grazing areas (frequently called range lands) by fencing, control of brush, stimulation or introduction of superior species with perhaps the application of fertilizers. The agronomists are also concerned with the establishment in appropriate environments of sown or planted grass/legume crops and their subsequent management by optimal control and management of the grazing animal, the so-called animal agronomy, for the maximum benefit of both sward and animal.

In order to discuss this "applied' work alongside data of a more fundamental nature for the benefit of vegetation scientists in general, we consider the place, the reasons for the occurrence of grass covers in natural vegetation, beginning with the ecological history and evolution of these grass covers, with their associated tree, shrub and herbaceous species, in response to natural and anthropogenic factors; the analysis of these covers according to botanical criteria, and the assessment of their value as an actual or potential economic resource; the nature of ecological regression and progression within the grass covers themselves, and in the vegetation of which they form a part.

What is frequently called range management is the correct integration of wild or domestic livestock with the plant species available for grazing or browsing in the natural vegetation. It is the response to ecological management of the plant species which is considered in this more botanical approach. The role of fire as an ecological/historical factor and as a management tool is of major importance throughout the zone of tropical grazing lands. The removal or replacement of natural covers by reseeding or planting raises questions of interspecific compatability or competition (in seeds mixtures), the nutritional and moisture requirements on soils of varying fertility (application of fertilizers and irrigation), and reaction to defoliation (in different grazing and cutting regimes).

1.3 Semantics

There is considerable disagreement on the correct nomenclature to apply to the different types of tropical grazing lands. The term 'savanna' is used throughout Latin America, by French, German and some British writers in Africa, particularly West Africa, and by some writers referring to certain types of forest/grass associations in Asia and Papua New Guinea. British and American workers in East Africa, however, have in general rejected the use of the term 'as being a South American one seldom correctly applied outside of its continent of origin' (PRATT, GREENWAY & GWYNNE, 1966). Some object to the use of a Spanish word of Caraib Indian origin for different physiognomic types of vegetation (see definitions in chapter 3.2.1); perhaps they forget that the American term 'range' is of late Middle English and French origin, used in the United States since 1766, to apply to 'grazing grounds' with physiognomy, botanical composition and ecological status completely different from tropical and subtropical savannas. The 1955 Yangambi Conference (Conseil scientifique pour l'Afrique au sud du Sahara, 1956) decided on a very restricted meaning for the term 'grassland' (grass = Old Aryan ghra – to grow; Old Teutonic; Old English; Common Teutonic), for which reason RATTRAY (1960) introduced the non-committal term of 'grass cover'.

BOURLIÈRE & HADLEY (1970) state: 'A major factor in classifying savannas is that the physiognomy of anthropogenic savannas is very often quite similar to that of natural ones. Human influence has undoubtedly extended the range of natural savannas so that savanna communities occur under many different climatic and edaphic conditions'.

KEAY (1959) has discussed the origins of derived savannas. The UNESCO meeting on forest borders (HILLS & RANDALL, 1968) concluded that savannas are 1. not a climatic climax, 2. the majority of units of this vegetation can be considered as anthropic, and 3. there are also natural savannas which can be considered as an edaphic climax (see references to AUBREVILLE papers in Chapter 3).

Recommendation no. 5 of the UNESCO Abidjan Symposium on tropical soils and vegetation may be quoted (UNESCO, 1961): 'Terminology for the classification of plant formation.

'In view of the serious difficulties, for all tropical vegetation students, resulting from the present confusion in vegetation terminology for the classification of plant formations, and with appreciation for the results of the meeting organized by CCTA/CSA at Yangambi in 1955, the Symposium invites the Director-General of UNESCO to convene a similar meeting of specialists on the vegetation of various tropical regions of the world in order to establish a project for a unified nomenclature of tropical plant formations on a worldwide scale, classified on the basis of the characters of the vegetation itself, rather than on those of its environment'.

In the present text, the term grassland is avoided whenever possible. Savanna is used when the original author(s) adopted that term – otherwise grass covers, tropical grazing lands and communities; range land in papers by American writers, or from East Africa, or when referring to the Rangeland Unit of the C.S.I.R.O. Division of Land Research in Australia, where the term is used to designate the arid and semi-arid areas unsuitable for crop production (partially tropical grazing lands), covering 74 per cent of the continent.

1.4 Succession and climax

The characteristics of regression under anthropogenous treatment, regeneration in response to environmental conditions and the effect of the biotic factor (grazing and browsing livestock) were obvious to the Chinese philosopher, MENCIUS, in the fourth century B. C.

孟子曰。牛山之木嘗美矣。以其郊於大國也斧斤伐之。可以為美乎。是其日夜之所息。雨露之所潤。非無萌櫱之生焉。牛羊又從而牧之。是以若彼濯濯。也人見其濯濯。也以為未嘗有林焉 (MENCIUS VI.A.8)

In the translation of WALEY (1937); 'The Bull Mountain was once covered with lovely trees. But it is near the capital of a great State. People came with their axes and choppers; they cut the woods down, and the mountain has lost its beauty. Yet even so, the day air and the night air came to it, rain and dew moistened it till here and there fresh sprouts began to grow. But soon cattle and sheep came along and browsed on them, and in the end the mountain became gaunt and bare, as it is now. And seeing it thus gaunt and bare people imagine that it was woodless from the start'.

It is currently fashionable in certain quarters to denigrate the significance of the concepts of succession (regression and progression) and of climax or subclimax in studies on grazing lands. Nevertheless, these terms appear frequently in the literature, and they belong to the vocabulary of field workers on tropical grazing lands. A University Department of Pasture Science in South Africa states (ROBERTS, 1970): 'The teaching and application of veld management leans heavily on the principles of plant succession, indicator plants and ecosystem dynamics . . .. Knowledge of potential climax condition is presupposed in the use of 'present condition' score sheets . . . for the assessment of condition based on five criteria, namely, cover density, botanical composition, vigour, soil surface condition, and insect damage'.

As a basis for further discussion, one might carry the proposition further, and state that the concept of succession, or something approaching it, applies at all levels of study: in types of vegetation, along climatic gradients, or by regression on the same site (MORELLO, 1968; 1.6); in types of grass covers, again along climatic gradients, or on the same site under the influence of anthropogenic and biotic factors; within genera of grasses, as in *Hyparrhenia*, for example, from those members of the genus high in succession to those at the lower levels (5.6), and within the species, from those ecotypes/phenotypes/genotypes adapted to one end of the humidity scale to those adapted to the opposite.

With movement in continual progress at these four levels and in opposite directions, is it a wonder that the taxonomists say that the members of the Gramineae are in a state of flux and active evolution? Unfortunately,following the excessive devegetation of recent and modern times, leading to progressive desiccation of habitats, the trend is all in one direction. Many of the superior genera, species and character groups within species will be lost.

Although it may be the ideal of the vegetation ecologist to maintain a climax flora of vegetation and grass cover, the top level in grass covers is not always the best objective of the practitioner. It was stated in the First Edition of WHYTE (1964) that, although the top level of grass succession is the best for arid and semi-arid Rajasthan, the top level in the high-humidity Terai (rainfall plus runoff from the Himalayan foothills) is of no economic value to livestock other than wildlife. The tall *Sorghum/Themeda/Narenga* grass stands of the Terai would have to be brought very low in the scale of succession before they would be of much value to domestic animals.

Due to the polarization discussed below in 1.5, it will be seen that the scientific bases for the improved management of tropical grazing lands are likely to become of increasing importance. The necessary botanical studies will include a much more fundamental analysis of the major component species of these grazing lands, the members of the Gramineae and of the main potential additives, the Leguminosae. In this connection, the ecological/historical analyses of the evolution of the major types of grass covers reveal a new outlook on the present taxonomic geography and genotypic composition and resources of grasses and legumes, and thus ultimately on their dynamic, rather than static taxonomy.

1.5 **Socio-economic factors**

Although the ecoclimatic and biological environment may be sufficiently similar in different parts of the intertropical zone to make a particular technique, e.g. the establishment of sown pastures, a feasible proposition in comparable rainfall belts, there are many dominant socio-economic factors which limit or inhibit such development. What is economically acceptable in the high-capital land economies of Australia is not applicable to the indigenous, low-capital communities of Asia and Africa. It then becomes necessary to propose realistic but cheap methods of improvement of grazing resources, e.g. by better management of free ranging livestock and a limited amount of surface seeding of legumes.

The conclusions of SPEDDING (1971) on the agricultural exploitation of natural grassland apply more to temperate ecoclimates than to the intertropical zone: 'The use of natural grasslands is often referred to as 'exploitation' because it is largely based on deriving production from land with the minimum input of resources. It is perfectly possible to raise the output of permanent grassland, by the use of fertilizers and by improved methods of utilization, to the level of output associated with well-managed leys'.

However, it is agreed that there are: 'several good reasons at present, and some which may remain for a long time, why some natural grassland will continue to be exploited without major inputs of additional resources . . . Where inputs are limited or the environment imposes severe restrictions, the herbage species adapted for survival may well be then considered less valuable from other points of view'.

The management of tropical grazing lands calls for the adjustment of livestock numbers (domesticated and wild), their seasons of grazing and their movements (nomadic or migratory) in such a way that the grazing lands are maintained at the most desirable botanical composition and stage of succession. In the drier areas, this may be achieved by appropriate location of stock watering facilities and by their periodical closure when the area they command begins to suffer through excessive pressure of livestock.

Due to the increasing awareness of certain economic factors, which differ in relation to the specific biological environment and the socio-economic structure, there is a marked tendency towards some degree of polarization in policy, research and practical advice on grassland and pasture matters. In most of the tropical zone outside Australia, it would appear that the provision of feed for livestock will be concentrated on the uncultivated and/or uncultivable land at one extreme, and on certain highly efficient areas of fodder cultivation at the other.

Three actual or possible situations may be mentioned, as variants of this polarization in tropical and borderline conditions. These, and many others which could be quoted, indicate the great importance of correct management of the tropical grazing lands, as one component of these economic systems.

India:

a. extensive free-range grazing for sheep, goats and camels, and for the rearing of draft cattle, breeds adapted to the seasonal availability of feed on the semi-arid grass/shrub communities on the uncultivable lands;

b. no sown pastures or leys; fodder crop acreage on cultivated land insignificant and likely to decrease due to demand for food for direct human consumption;

c. village cattle fed on crop residues and concentrates;

d. cultivation of gross-feeding African grasses on sewage farms, cattle colonies and other areas where a very high level of plant nutrients and water can be maintained; green feed (supplemented with concentrates) for high-producing milch cows and buffaloes; economic return per unit area in terms of an ultimate livestock product such as milk higher than can be obtained from any other crop.

Japan:

a. sown pastures on mountains and sloping land too steep for crops, for grazing in the growing season by dairy heifers and young beef animals;

b. no sown pastures or leys and insignificant area of fodder crops in the cultivated areas (summer monsoon conditions);

c. stall or yard-feeding of productive milch cattle in zero grazing systems providing green or conserved feed, plus concentrates;

d. feeding of cattle for beef production and of growing stock from hills (in winter) in feed lots with locally produced feeds and imported grain.

Kenya:

a. free-range grazing of unimproved Boran cattle on the semi-arid lands;

b. fattening in special feed lots, using maize grain (1 per cent of total national production) and/or silage, and molasses as a by-product of the sugar-growing industry around Lake Victoria (UNDP/FAO Project).

See bibliography no. 1248 of the Commonwealth Bureau of Pastures and Field Crops for 149 references to literature on grazing systems in the tropics and subtropics.

1.6 **Ecosystems**

Two Canadian specialists (HARE & RITCHIE, 1972) writing on a non-tropical environment express the modern outlook succinctly: 'Present-day ecological research is concerned, among other things, with the productivity of ecosystems (that is, the energy cycle of the biosphere), with ecosystem dynamics or their capacity to evolve as complex systems, with populations dynamics, with biogeographical history, and with the structure of natural vegetation, including the role of animals in maintaining it'.

The natural and the cultivated or synthetic herbage communities and their constituent species discussed in these chapters are the initial points in biologic and economic ecosystems which lead ultimately to the

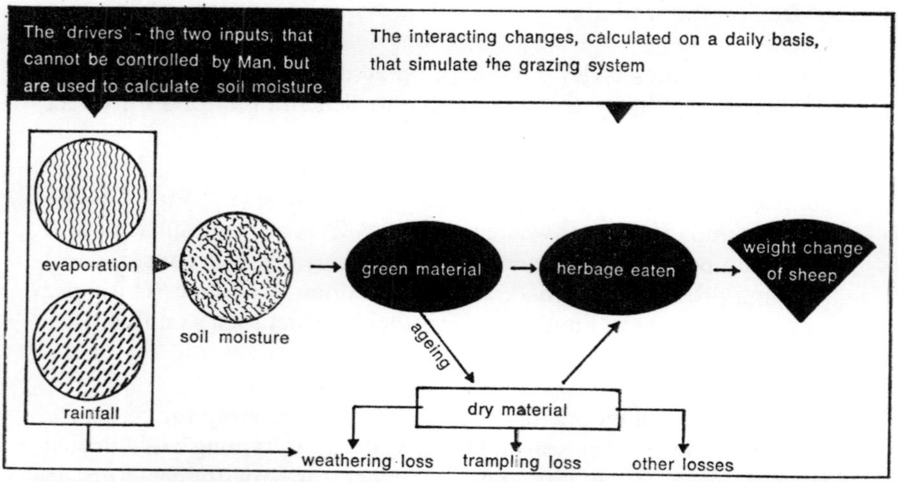

Fig. 1/1a: Mathematical model of grazing system evolved in Australia (*Rural Research in CSIRO* 75, March, 1972). This diagram shows the components of the first mathematical model built at Canberra to simulate the grazing of summer pasture by sheep. It calculates daily changes in the amount and digestibility of the food available and the diet eaten, and the resulting changes in the weight of the sheep. A more ambitious model is being attempted to predict year-round meat production from both cattle and sheep.

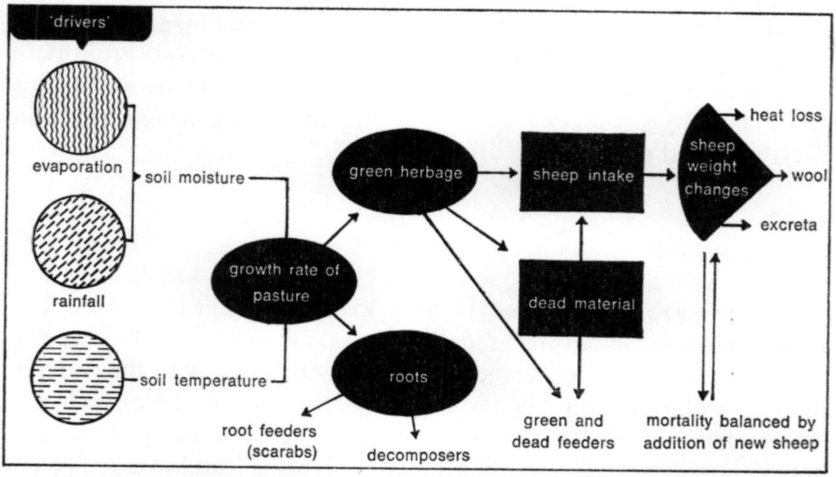

Fig. 1/1b: Mathematical model of grazing system evolved in Australia (*Rural Research in CSIRO* 75, March, 1972). The Armidale group adds soil temperature as an extra 'driver'. Greater attention is paid to losses from invertebrates such as scarabs, millipedes and nematodes. These losses, together with heat production, wool growth, excreta and mortality balance the total energy input.

provision of a livestock product of value to man. Most of the ecosystems in the tropics which have a livestock product as the ultimate objective are unbalanced or in disequilibrium, using these terms to apply to ecosystems in which one or more of the major constituents, plant, animal or man, have a chronic or seasonal lack of nutrients, water or other environmental requirement. It is the objective of the practitioners in the land sciences to overcome these deficiencies to the maximum extent possible, and of the vegetation scientists to make all possible contribution within the scope of their own disciplines. But it is the grazier and the farmer, the managers of the ecosystem, who have to decide whether the proposals of the scientists are economic and acceptable. Most animal husbandmen in the intertropical zone are part of the land ecosystems with their domestic livestock, especially in semi-desert and nomadic communities. A relatively few live outside the land ecosystem which they operate, in outlying domestic 'islands' approaching the standards of urban ecosystems.

The objective of ecology is to study variations of primary production of plant communities, natural or synthetic, in the different climatic zones of the biosphere (WALTER, 1971a, MISRA, 1969). GILLARD (1969) proposes 'ecosphere' as an alternative to J. B. LAMARCK's term (defined as the rough total of the matter of all living beings); ecosphere being that part of our sphere, in which there is life together with the living organisms it contains. Just as the biosphere is the rough total of all living organisms, so ecosphere is the rough total of all ecosystems. MORELLO (1968) considers that there is a gradient of complexity and size of ecosystems (again the concept of succession) and that the 'great unit' (gran unidad de vegetación y ambiente) as an ecosystem is on the top of that gradient. 'The entire area occupied by one model repeated many times is a great unit of vegetation and environment.'

It will be necessary to consider how the mathematical model/computer technique may be applied to the still rather primitive tropical ecosystems (MACFADYEN, 1964; MORLEY & SPEDDING, 1968). A review of Australian work on the construction of mathematical models of grazing systems appears in Rural Research in C.S.I.R.O., No. 75, 1972 (Fig. 1/1). Most writers on this subject state how to design a model, but not how to apply it under field conditions (SPEDDING, 1971, Appendix). 'The complexity of grassland ecosystems imposes an interdisciplinary team approach for research and management. We must now train many grassland scientists and managers to work in interdisciplinary teams' (VAN DYNE, 1970). LUGO (1969) discusses energy flow in some tropical ecosystems (rain forest, mangrove swamp, tropical savanna and slash- and-burn agriculture). The systems approach adopted by MOTT, EDDLEMAN & TIMM (1970) for a study of the soil/plant/animal complex in the tropics is based on cultivated fodders. Care must be taken to avoid a situation where students in less advanced countries become distracted from the

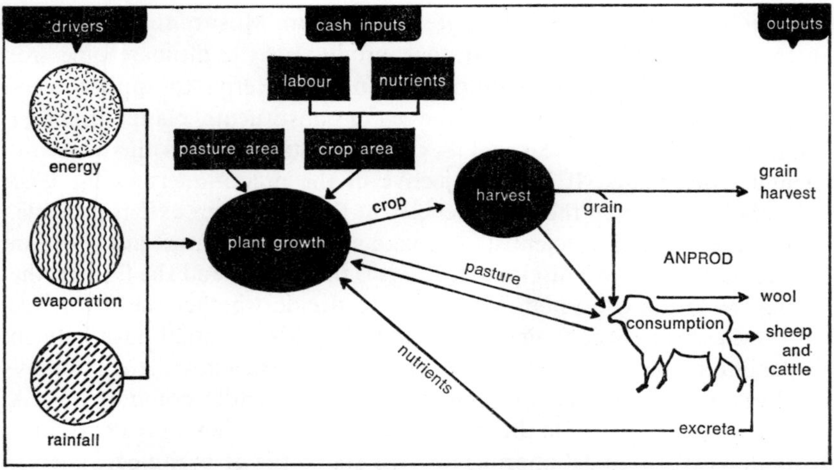

Fig. 1/1c: Mathematical model of grazing system evolved in Australia (*Rural Research in CSIRO* 75, March, 1972). A simplified version of the Perth group's ley farming model, which introduces crops as well as pasture and simulates both cattle and sheep growth. So far only part of it, ANPROD, has been tested. Another sub-routine, still to be built, will deal with reproduction of livestock.

practical problems of their country by the model-cum-computer philosophy, developed largely as a theoretical exercise in academic university departments and research institutes in the more advanced countries.

Space does not permit a review of the ecosystem concept as it relates to all the diverse ecosystems of the intertropical zone in terms which can be appreciated and applied by tropical field specialists, for most of whom English is not their native tongue. This is a task for the future (McMillan, 1969; Dix & Beidleman, 1969). A start has been made at the meeting in Côte d'Ivoire from 29 December 1971 to 3 January, 1972, organized under the auspices of the International Committee of the I.B.P. Grasslands section. Papers related to studies in the Miombo Project, Serengeti Research Institute, the Lamto Project, to biomass and primary productivity in various types of savannas, and other aspects of the research in Africa discussed in the following Chapters (over thirty papers being published in the Annals of the University of Abidjan).

The I.B.P. Grassland group has also considered approaches to the synthesis of results. 'Synthesis is a necessary step in advancement of the science of ecosystem study and final understanding of ecosystem functions. It is a kind of inductive reasoning from particulars to generalization. This is a task of making the whole greater than the sum of the parts and of comprehending the interrelationships of system-subsystem properties', in word models, comparative analysis, subsystem models and total-system models (I.B.P./PT Grasslands Newsletter No. 7). The requirements for total-system models are illustrated in Fig. 1/2 and Table 1/1.

Table 1/1

Kinds of synthesis proposed by the IBP Grassland group (see Fig. 1/2)

Kind	Primary Type of Data	Amount of Raw Data Required	Scientific Usefulness of Output (For Whom)[a]	Utilitarian Usefulness of Output	Size of Audience Potential	Size of Audience Present	Advantages	Disadvantages
Word models	All	Low	All	Low	Small	Large	Simple	
Comparative analysis	Driving variables State variables Site descriptors	Medium	Project National International	Medium	Medium	Large	Usefulness in model testing	
Subsystem models	Rate processes (one-few)	Medium-high	Subproject International	Medium	Medium	Medium	More international comparisons possible	
Total-system models	Rate processes (few-many)	High	National International	High	Large	Small	Basis of futuristic renewable resource management and 'interpretation of subsystem models'	Few data information sets available

[a]) Subproject = research of an individual scientist.
Project = research of a group of scientists.
National program = research on a group of projects.

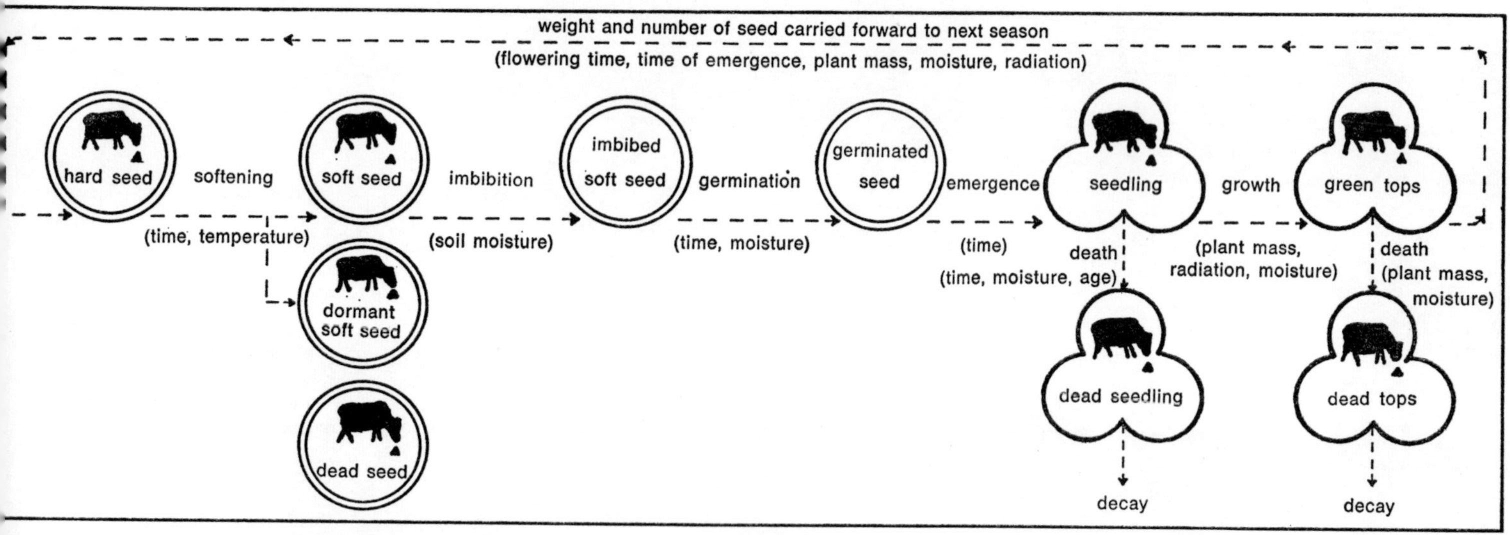

Fig. 1/1d: Mathematical model of grazing system evolved in Australia (*Rural Research in CSIRO* 75, March, 1972). ANPAST, one part of the Perth group's ley farming model, simulates the yearly growth of an annual pasture. It calculates each stage of seed development and emergence, plant growth, and seed set. ANPAST measures not only the effects of moisture and sunshine on these processes, but also those of stock grazing at various stages in the cycle.

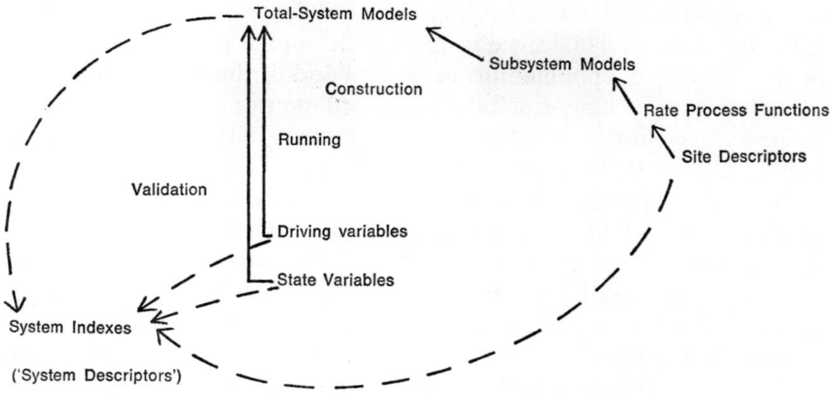

Fig. 1/2: Total-system models evolved for the IBP Grassland group.

The International Coordinating Council of Man and the Biosphere (Secretariat in Division of Natural Resources, UNESCO) is interested in the impact of human activities and land-use practices on, inter alia, the savannas and related tropical grazing lands, in the following respects:

(i) studies of the effects of land management and grazing on the transformation of ecosystems, including changes in bioclimatic conditions;

(ii) comparative studies of domestic and wild animals (as individual species and as natural groups, within ecosystems) with a view to finding the most efficient producers of protein; where appropriate as tourist attractions and natural gene pools; and as influence in the maintenance of integrity and stability of ecosystems;

(iii) studies of the effect of fire in different areas, including the co-ordination and more scientific pursuit of existing experiments on different intensities and times of burning.

Included in studies of tropical ecosystems should be:

a. human anthropology in relation to type of tropical grazing lands and the animal husbandry maintained thereon;
b. wildlife on tropical grazing lands: feeding habits, use of fire (TALBOT, LEDGER & PAINE, 1961; LEMON, 1968b; FIELD, 1970; RINEY, 1970; BOURLIÈRE & HADLEY, 1970; ROSS, 1969; CASEBEER & KOSS, 1970; FAO/UNDP, 1971)(also F. KLÖTZLI's forthcoming study of the biogenous influence on natural vegetation from the montane to the afroalpine belt in northern Ethiopia).
c. the pasture/coconut ecosystem, in South Pacific Commission Territories (HUGH, 1968); Papua and New Guinea, and Sri Lanka.
d. parallelism between the ecologies of plants and domestic animals; the level of feed availability governing the standard of animal husbandry; the integration of different types of fodder availability and production, to provide sustained yield of animal products.

For example, land ecosystems designed to produce milk economically are of great significance in raising the animal protein components of diets of tropical populations in general and of the vulnerable groups of society in particular, especially in western monsoon Asia. The tropical pasture grass constituents of such ecosystems vary in their milk-producing ability. C.S.I.R.O. Division of Tropical Pastures finds that, although voluntary intake of tropical legumes is usually higher than of tropical grasses of similar digestibility, less milk is produced from *Phaseolus atropurpureus* cv. Siratro and *Desmodium intortum* c.v. Greenleaf than from *Digitaria decumbens* fertilized with nitrogen. The composition of milk is affected by the poor quality of tropical feeds; the cow must draw heavily upon body resources. In a trial on *Stylosanthes humilis/Heteropogon contortus* pasture (RITSON, EDYE & ROBINSON, 1971), cow conception rate was significantly and positively correlated with the phosphorus contents of the three major components (stylo, perennial grasses, annual grasses), with the sulphur content of the stylo, and the three yield attributes of annual grasses (dry matter, phosphorus and nitrogen). The following productivity may be achieved on a tropical pasture of *Leucaena leucocephala/Panicum maximum*, from nine months' unsupplemented grazing from six cows (in kg. per ha.): milk 1650, fat 245, protein 190. The climatic limitations of existing tropical pasture legumes are discussed in 10.6.

1.7 **Wildlife in African ecosystems**

It is in East Africa in particular that the end product of management systems on tropical grazing lands is wildlife, whether produced as a source of protein in the human diet, as a tourist attraction bringing in foreign exchange, or for other purposes.

Meat of all species of wild ungulates is eaten readily by various tribes in East Africa (TALBOT, LEDGER & PAYNE, 1961); many species such as eland and Thomson's gazelle are also prized by Europeans. Since most meat except that of zebra contains little fat, it can easily be converted to biltong or other dried meat products where ready outlets for fresh meat are not available. There are good markets for hides, either locally or overseas, and for other animal products such as elephant ivory, rhino horn, hippopotamus teeth, giraffe and wildebeest tails (for fly whisks).

The standing crop of wild animals on *Acacia* savanna land may be two to eight times that of domestic livestock, and the crop of wildlife on bushland may be from four to fifteen times that of goats and sheep (TALBOT *et al., op. cit.*).

'Since the wild animals breed earlier, reach maximum growth quicker, and achieve higher killing-out percentages than the domestic live-stock

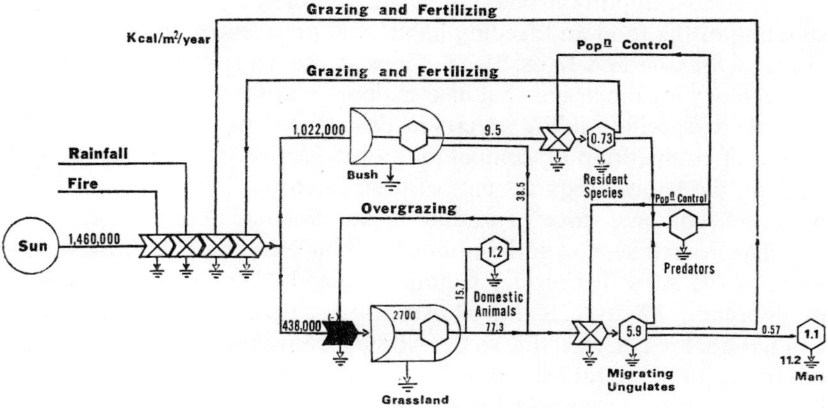

Fig. 1/3: Energy diagram for the Serengeti Mara Plains Region, Tanzania, based on data of TALBOT & TALBOT. The caloric value of animal biomass was assumed to be 1.4 Kcal/gm, wet weight. The caloric requirement of man was assumed to be 2,000 Kcal/day. Grazing acts as a positive feedback by maintaining the vegetation in a young productive state. Overgrazing acts as a negative feedback since it represents an exploitation of the vegetation leaving a low amount of photosynthetic tissue for recovery to occur. If only the feeding area is considered, the standing crop of ungulates may be 22 Kcal/m². However, this value neglects the total area of the home range which becomes very important during periods of moisture stress (LUGO, 1969).

on comparable land, it can be assumed that, other things being equal, the potential harvest of protein and other animal products from wild animals should be several times that possible from domestic livestock alone on much East African rangeland.'

Wise manipulation of control and use of wild animals requires a better understanding of their role in the ecosystem and the energy flow stream (FAO/UNDP, 1971). An energy diagram and suggested work flows proposed by LUGO (1969) are reproduced as Figs. 1/3 and 1/4.

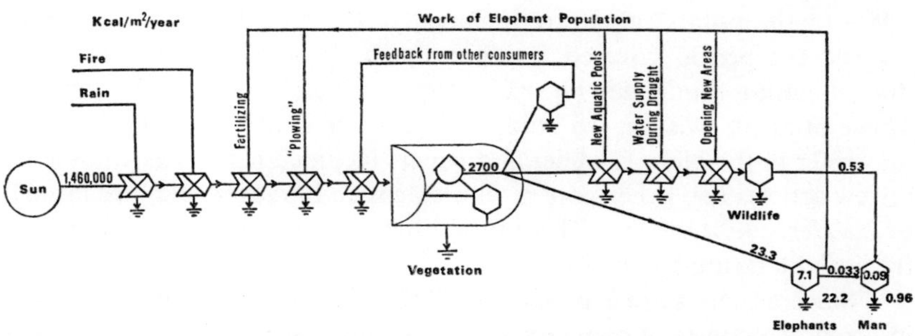

Fig. 1/4: Suggested work flows of elephant populations in the Luangwa Valley, Zambia. Wildlife and vegetation production were assumed to be equal to those reported in Fig. 1/3 (LUGO, 1969).

The most important phase of evaluation of ecosystems is an understanding of the food and feeding habits and preferences of the occupant fauna (CASEBEER and KOSS, 1970). Gaps in the fields of wildlife biology and ecology on rangeland that also maintains domestic animals include the role of specific wildlife species in disease and parasite relations, the degree of competition or compatability between domestic livestock and wildlife, the habitat requirements of designated wildlife species, and the potential economic value of wildlife in situations of multiple use.

The Fourth Session of the ad-hoc Working Party on Wildlife Management of the African Forestry Commission of FAO was held in Nairobi in February, 1972. K. CURRY-LINDAHL discussed ecosystems, biomes and national parks, referring to the African Convention for Conservation of Nature and Natural Resources, to the fact that most national parks are located in areas unsuitable for development, to the threats to existing national parks, and to the urgent need to conserve the components of the biomes, including the ecosystems comprising the grazing and browsing resources. For these purposes it is necessary to consider the place of wildlife in land-use planning (W. G. SWANK and R. L. CASEBEER, UNDP/FAO, Kenya), and the problems associated with harvesting and marketing the wildlife 'crop' (P. THRESHER). Any such developments will in due time increase the economic resources available for improving the initial stages in these ecosystems, that is, the condition of the rangeland itself.

1.8 Coverage

The present study is not a review of the literature. Rather is it a consideration of some of the main scientific principles and objectives in research which can be seen in operation, or apparently arising at the present day. In certain aspects, a personal bias may be seen, especially where heterodox propositions might lead to profitable discussion and research.

Only those published papers and personal communications that relate to the points raised are included in the already extensive bibliography; the period covered is primarily the past five to ten years. No discrimination is intended regarding papers which have had to be omitted. Those students wishing to study any particular aspect in more detail may refer to the bibliographies and papers included here (where possible, review articles and other papers with extensive bibliographies are listed), or to *Herbage Abstracts*. This journal has covered all tropical and subtropical literature for many years.

Sincere thanks are expressed to all those who assisted so generously, in providing copies of congress and conference reports, books and occasional publications, reprints, documents of U.N. Agencies, and personal communications.

1.9 Contrasts between tropical and temperate latitudes

The scientist who has received his training and early experience on temperate grazing lands and pastures, and who comes to the intertropical zone in programmes of technical assistance, is faced with a set of entirely different conditions. The subsequent chapters of this book have been written primarily for the benefit of workers already in the tropics, by one whose major interests have been in the tropics. It may be appropriate at this stage to refer briefly to some of the aspects which will be unfamiliar to the newly-arrived specialist from temperate latitudes.

Tropical grazing lands in their natural, unimproved condition are composed of a great number of species of Gramineae, which present difficulties in recognition, taxonomy, ecology and genetic composition and behaviour. Some grasses have achieved pantropical distribution without necessarily being good pasture plants. Species of the Leguminosae, so important in temperate pastures, do not acquire great significance in the tropics until the higher levels of intensification are reached. Shrubs and trees are essential components of tropical grazing lands, especially of the more arid zones.

The contrasts between wet and dry seasons are much greater over vast areas than in the temperate regions. After the original vegetation has been destroyed, the communities in the secondary stages of succession are exposed to increasing or recurrent changes in the habitat, towards greater aridity, induced mainly by man and his domestic animals on the free range. Fire has played, and continues to play, a significant role in the origin, botanical composition and management of tropical grazing lands.

As the grass covers over wide regions of the tropical grazing lands are still relatively young and not yet stabilized, as in Europe, existing as it were in a constant state of flux, the capacity to interpret climax and plant succession becomes much more important. Provided seed-producing plants of species higher in succession still exist, promotion of succession upwards may be a rapid and efficacious method for initial management, following reduction of the biotic factors associated with excessive or badly-timed grazing pressure.

Perhaps above all, the newcomer comes to accept that enough is known about most types of tropical grazing lands for correct advice to be given on measures for management and improvement, but that their application in the field is limited, if not actually prevented, by overriding social and economic factors. At the sociological level, one has first to consider the distinction between the cultivator and the pastoralist, and the low technological level and poverty of the latter. Many traditional customs and prejudices are associated with the different types of livestock and their husbandry (capital on the hoof, bride price, sanctity of bovines, etc.).

Table 1/2

Gradients of increasing availability of soil moisture, soil fertility and intensification of feed and livestock production

Resource	System of use	Type of livestock	End product
1. Unimproved range, mostly arid and semi-arid	Free range grazing, mostly nomadic and/or migratory	Cattle, sheep, goats, camels, wild animals, adapted to wide seasonal fluctuations in supply of feed and water	Beef, meat, wool, hair, hides, skins, wildlife products, tourism
2. Improved range, semi-arid, or savannas in humid tropics	Numbers and movement of stock controlled by herding, or by closure of watering points	Similar	Similar
3. Dryland fodder production, crop residues and stubble grazing	Stubble grazing in dry season in a monsoonal environment, with yard feeding or free-range grazing in the rainy (cropping) season	Slight improvement in quality of livestock, in line with slight improvement in total and seasonal availability of feed	Specially relevant to cattle populations in villages of India, (25% of world's total) giving minute amounts of milk in a four-month lactation
4. Sown or planted synthetic stands for grazing	Rotational or fixed grazing during growing season in fenced pastures, other arrangements for dry season	Change of type of livestock and introduction of superior breeds justified, especially milch animals or cattle for fattening	Milk, meat or beef according to social acceptability
5. Semi-intensive fodder production on cultivated land	Cut and carry, zero-grazing system for crops such as lucerne, berseem or grasses growing alone or with legumes	Production from superior animals ensured by availability of high-protein cut green fodder, or hay or silage	Similar
6. Intensive fodder production	Gross-feeding African grasses grown with irrigation water plus heavy dressings of nitrogen or with cowshed wash or with treated city sewage	Ample green feed, requiring supplementation by concentrates, for productive milch animals in specialist dairy units or milk colonies	Milk
7. Import of feeds, fodders, concentrates or coarse grains from elsewhere in same country or from abroad	Feeding as supplements to roughages and cut green fodder, in stalls, yards, or feedlots	(a) Productivity of high quality milch cows and buffaloes assured (b) Cattle brought in from range or mountain pastures for fattening	(a) Milk (b) Beef

Some of the economic problems have been discussed under 1.5. The major ones facing the would-be improver of the resource and of the types of animal husbandry practised thereon, are the vast extent and low carrying capacity (and therefore low potential return on investment) of the grazing lands, lack of funds sufficient to cover such an area, and the problems of extension among such scattered, sparse and generally moving populations. Accessibility of the grazing lands to markets and to centres of health and education is also significant. Four levels of investment and production can be recognized, providing for a progressively more reliable seasonal availability of better quality plant material (6.2).

The relative composition of economic ecosystems and of the different levels or plateaux of production on a scale of increasing intensity of utilization are set out in Table 1/2. Decisions regarding the technical and economic practicability of raising an economic ecosystem from one level to the next higher, or of combining two or more ecosystems in an integrated productive system, may be made on the basis of the special economic criteria that apply to the tropical situation. The higher the level of production, the more costly is the feed. The areas involved are so vast, the climatic limitations so severe and widespread, and the great livestock populations of such low productive potential, that concentration of effort on relatively small and favourable areas and enterprises must surely be economically correct.

2 EVOLUTION OF GRASS COVERS: AUSTRALIA AND ASIA

2.1 Technique of ecological/historical analysis

If ecological analysis is combined with a study of the history of vegetation and devegetation, it is possible to say which type of grass covers, communities or species could or could not have been found in the natural vegetation as it existed at any particular period of geobotanical history, and especially before the advent of man. The primitive grass species of forest shade are neither adapted nor sufficiently aggressive to spread with the clearance of forest and the increase in light-intensity – these new areas must have been colonized by light-loving, more xerophytic grasses from elsewhere. It is necessary to turn to the history of human migrations and trade in order to find an explanation for the present composition of many if not most of the grass covers of the intertropical zone.

Specialists in the taxonomic geography of the Gramineae have, in association with forest ecologists, to learn to categorize forest covers in respect of their capacity to support growth and reproduction of particular members of the Gramineae. The major factor is, of course, shade. For example, the criteria developed for south-east Asian tropical rain forest (2.3) probably apply to tropical rainforest everywhere; also to other types of forêts denses, until one reaches a type of forêt claire with sufficient light at ground level to meet the needs of light-requiring species.

The ecological niches in tropical rainforest in south-east Asia in which one might expect to find, or to have found, the few species of the Gramineae adapted to such conditions are:

- within the forest itself, or on its fringes
- coastal sands and muds
- river and stream banks in the interior
- high altitudes
- areas of tropical rain forest disturbed not by man but by severe windfalls in steep terrain with unstable soils (ANDERSON, 1971), or by slips related to long periods of heavy rain, sites that may take many decades to be colonized by anything but grass and ferns (BURGESS, 1972)
- salt licks and grazed areas in the forest, especially in Borneo, visited by guar and banteng (BURGESS, *op. cit.*)
- areas affected by fire (cooking fires of hunters), or by the severe windstorms of 1883 in Kelantan, said to be related to the eruption of Krakatau and possibly responsible for the 'very odd size-class distribution in the 'storm forest' of Kelantan' (BURGESS, *op. cit.*).

The results of ecological and historical analyses are essential for a better appreciation of the present botanical composition of grass covers, also in the study of their generic and specific diversity (rich in tree savannas or comparable types of ancient grass covers, poor in relatively young derived savannas), and in the choice of areas for collection of grass genotypes (the desired variability having some relation to distance in time and space from centre of origin, and to the possibility of renewed speciation in an environment which exposes immigrants to physiological stress).

2.2 **Australia**

Long-term research on the naturalized plants in the native vegetation of Australia has started in the Department of Biogeography and Geomorphology at the Research School of Pacific Studies in the Australian National University, Canberra (WACE, 1971). Three main aspects are envisaged at this stage: 1. historical inventory of naturalized species. 2. biology of some naturalized species. 3. native vegetation and alien establishment.

This project will inevitably and in due course encompass the Gramineae of northern tropical Australia and thus become contiguous to the studies of monsoonal and equatorial Asia discussed below. It also raises larger issues concerning 'the stability of climax ecosystems, the opportunities for movement of vegetation, and the whole nature of plant migration before the massive disturbances of ecological equilibrium brought about by the spread of man . . . and his continual creation of fresh habitats for exploitation by plants' (WACE, *op. cit.*).

One example is that reported in Rural Research in C.S.I.R.O., No. 74 1971. Travellers in central and northern Queensland around the years 1840–1855 made no mention of *Heteropogon contortus*, now so widespread. The landscape must so recently have consisted of open woodland with an understorey dominated by *Themeda australis*.

Linking studies between the two regions of Australia and Asia include the observations on the anthropogenic grasslands in Sepik Plains, New Guinea (HAANTJENS, MABBUTT & PULLEN, 1965), and the anthropological study of ecological succession in ascending grassland in a high mountain valley in New Guinea (BOWERS, 1969).

2.3 **Eastern Monsoon Asia**

Subsequent studies will be directed towards an ecological/historical analysis of the grass communities and their constituent species in eastern monsoon Asia. The grass species found in the deforested monsoonal lands of China and in the transition zones between these and the Asian continental grasslands to the north and west have been listed by WHYTE (1968a, p. 215); a further analysis of distribution and habitats given in the flora of KENG YI-LI (1965) is in progress. NODA (1972) includes the Gramineae in his flora of Manchuria; many have subtropical or monsoonal affinities and are probably relatively recent arrivals in that environment. LEE (1966) has published a manual of Korean grasses. Japan has perhaps three genera (*Miscanthus*, *Sasa* and *Zoysia*) that may be regarded as truly indigenous, from the time when most of the country was covered with broad-leaved deciduous forest (probably grass-less); all other grasses found now are escapes of European or other exotic species, subtropical and temperate, which were introduced by accident or intent during the past 200 or 300 years. LIU (1970) has noted the grass-dominated consocies found in the following vegetation formations in Taiwan (for alpine tundra and coniferous forest formations, see LIU, 1971a, and also Fig. 2/1 for stages in evolution of the vegetation of Taiwan after the Late Cretaceous, LIU, 1971b):

a. warm-temperate rain forest (as in south-west China), with *Miscanthus sinensis*, *Indocalamus niitakayamensis*, certain bamboos and *Pseudosasa* in secondary succession,
b. tropical rain forest formation (Malaysian flora as in Yunnan and Hainan), with *Imperata cylindrica* var. *koenigii* and *Miscanthus sinensis* in secondary succession,
c. littoral forest formation,
d. tropical savanna formation, with the dominant grasses *Miscanthus sinensis*, *Imperata cylindrica* var. *koenigii* and *Rynchospora chinensis*; LIU finds it difficult to regard the tropical savanna of Taiwan as a climatic climax.

From an analysis of studies such as these, in relation to the history of devegetation of the region from former climax formations, it will become possible to separate the grasses into the indigenous and the historically irrelevant, as has been done for south-east Asia.

2.4 Mainland and insular South-east Asia

Before man's arrival in south-east Asia, the natural vegetation was a climax cover of various types of tropical rainforest, highland and swamp forest vegetation in mountains, valleys, plains and deltas (RICHARDS, 1961). There would have been no grassland climax nor climax grass covers in the region. Primitive species may have contributed in a minor way to the plant associations of the forest fringes and of open sites along river beds and in mountains. In terms of geobotanical history and evo-

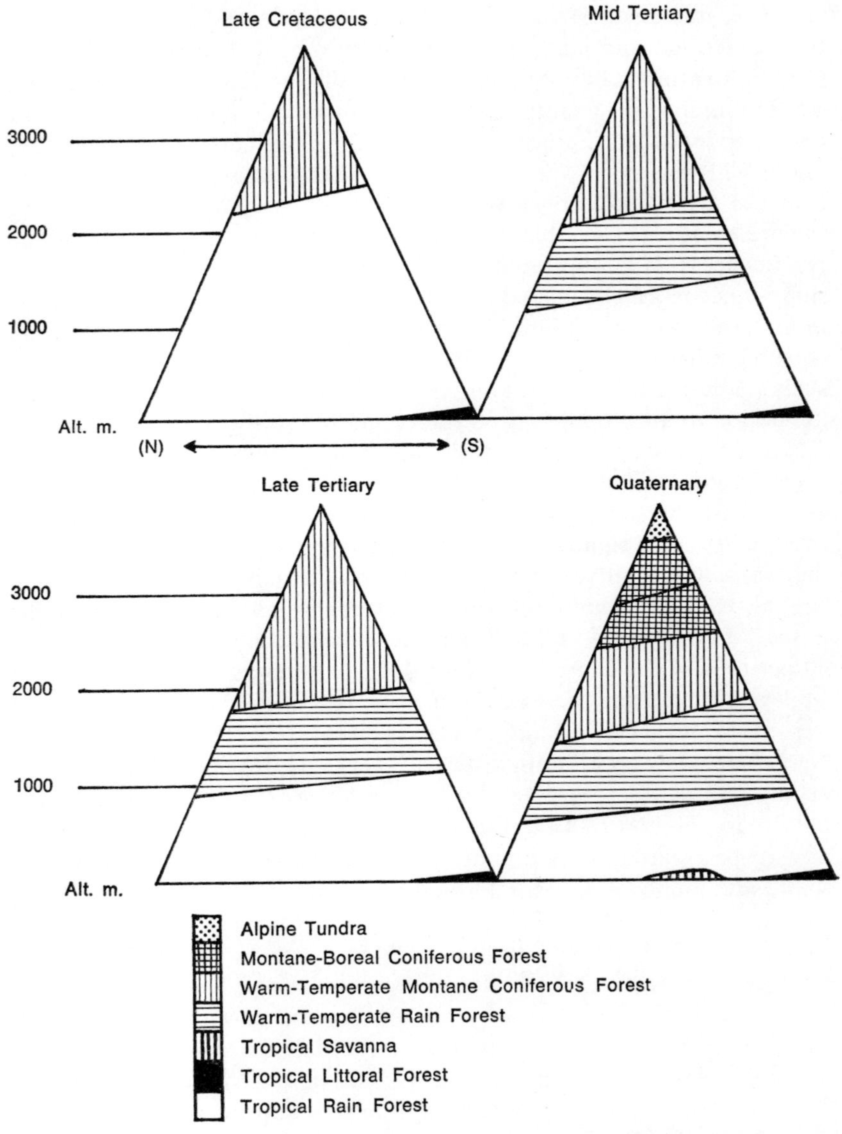

Fig. 2/1: Stages in evolution of the vegetation of Taiwan after Late Cretaceous (LIU, 1971b).

lution, man in any numbers is a very recent arrival indeed in the forest ecosystems of equatorial and subequatorial monsoonal Asia; his influence in changing the environment has been significant only in recent centuries (Papua and New Guinea/UNESCO, 1960; PELZER, 1968).

The grasses of Malaya (now West Malaysia and Singapore) have been described and classified taxonomically by GILLILAND and others (1971). An attempt has been made to reclassify the 220 species on the basis of their individual adaptation to environment as indicated by site of collection, or of their known environmental requirements (WHYTE, 1972b; 9.5.2). This has been done to separate those grasses which can on the evidence available. be regarded as truly or possibly indigenous pending the presentation of reliable data on habitat requirements and taxonomic primitiveness. It is only such grasses that are of sufficient geobotanical significance to be considered in fundamental studies of the origin of the Gramineae, early speciation, the nature of primitiveness, centres of origin of subfamilies, and related matters. All other species are irrelevant to such studies; they are, however, most important in studies of dispersal in relation to anthropology and trade, and in investigations of the possibility of renewed speciation in the secondary habitats which have now been created.

In this connection, it will be necessary for taxonomists to define the relative historical significance of endemic species (9.5.3); of species which they consider primitive because of their floral structure (9.2.2); and of the species which are found and which presumably have long existed in the shade of tropical forests. It has been proposed (9.5.2) that up to 80 per cent of the genera now recorded for Java, Sumatra, Borneo and Malaysia came from the west through human agency – from Africa and western Monsoon Asia – and to a lesser extent from the east – Papua and New Guinea and northern Australia – very late in the geobotanical calendar. This fact is recognized for Java by MONOD DE FROIDEVILLE (1968) but not for Malaya by GILLILAND and others (1971). Many of the species now to be regarded as newcomers in south-east Asia, mainland and insular, are found in the coral islands and atolls of the Indian Ocean (STODDART & YONGE, 1971) and in the Andaman Islands (BHATTEE & THAMPI, 1963). WILLIS & GARDINER (1931) record twelve species on Chagos Archipelago; SUMMERHAYES (1931) records for the Seychelles one endemic and 42 species of pantropical distribution, or with African, Mascarene or Asian affinities. These records need to be analyzed in relation to the age of the coral islands and atolls, the availability of watering points for early mariners under sail, the movement of bird populations, and the grass floras of east Africa, Arabia and western Monsoon Asia.

The natural vegetation of the Malay Peninsula is High Rain Forest 'holding the most intense rain-forest flora of the old world' (BURKILL & HOLTTUM, 1923/5), which only some one or two centuries ago (CHENG, 1969) dominated all but the maritime fringe, river plains, occasional forest clearings and exposed mountain ridges. Even now this type of forest covers more than half the land area, although it is being further encroached upon by land development schemes such as the Jengka

Triangle and the similar project in Johore. Rapidly changing ecological conditions will now lead to the progressive discovery of species hitherto considered exotic on sites opened to the influence of intense light and on soils which have become more arid, with the structure of gramineous or cultivated soils. One example of this has been studied by VERBOOM (1968) in Pahang, central Malaysia (5.9.2).

Many events in the anthropology of the Asian region would have contributed to the adventitious introduction of seeds of grasses from far distant countries. They came in packing materials, as weed seed contaminants in grain supplies, or in the sweepings of godowns or of the decks on which cattle travelled. Where the new environment was suitable, these species became established around coastal settlements and along riverine routes into the interior. The great expansion of the exotics followed the opening of the countries of Malesia for plantation agriculture, which began only in the 18th century and became widespread only in the 19th. Stands of grasses grew up around the houses of plantation workers and in their food plots, where the combination of higher light intensity and drier soils provided conditions necessary for non-indigenous species to thrive.

Of great significance to the spread of these grasses to the foothills and more remote areas of Malesia was the movement of the shifting cultivators. They carried uncleaned crop seeds with their weed seeds wherever they moved. For much of the history of Malesia, regeneration of the forest cover was so rapid that these exotic herbs would have been eliminated in the long forest cycle between burnings. They could have been renewed only by repeated contact with people from permanent settlements (silent barter as a method of accidental seed introduction into remote mountain clearings, BLAGDEN & SKEAT, 1906). During the past hundred years, however, pressure on the land has increased, the shifting cultivation cycle has become ever shorter, preventing regeneration of secondary forest and creating conditions whereby the herbaceous cover has become permanent, and dominated by the ubiquitous species, *Imperata cylindrica.*

Thus, the grassland ecologist and agronomist concerned with the vast expanses of grass covers of various types, although predominantly *Imperata cylindrica* in Indonesia and the Philippines in particular, should become aware of their relative youth in terms of geobotanical time. MERRILL (1906) made such an analysis for the Gramineae of the Philippines. He recognized 225 species and varieties; few are constituents of the *Imperata* grasslands, most occurring in specialized localities such as river banks, seashores, forest floors, roadsides, etc. MERRILL classified his genera and species into intentional or accidental introductions, endemic, cosmopolitan-tropical, Australian and Malayan.

'On the whole, the Philippine Gramineae are strongly Malayan or

Indo-Malayan, with a decided northern element in the highlands of northern Luzon, and a rather characteristic Australian one, which, strangely, is from the same northern region of the Archipelago, rather than from the southern islands, although when more extensive collections have been made in the interior of Mindanao, doubtless most of these Australian types, which are at present known in the Philippines only from northern Luzon, will be found in Mindanao.'

In Malaya and Java in particular, pending further analyses, it is possible to recognize three periods or stages of increasing intensity of anthropogenic influence (see 9.4):

(i) the millenia before man,

(ii) from the appearance of man in small numbers up to relatively recent times, and

(iii) the present, beginning in what is now Malaysia only about 100 years ago, and in Java about 200 years ago, when large-scale clearance of forests created and is still providing extensive areas of land characterized by the absence of shade and of forest humidity, and by more arid soils, an environment suitable for non-indigenous light-requiring grasses, which must necessarily have arrived from elsewhere.

During much of the prehistoric and most of the early historical times, the hinterlands of East Africa and much of western Monsoon Asia were grasslands or savannas. In Malesia, and in south China as part of eastern Monsoon Asia (WANG CHI-WU, 1961), they were forest, with very few grasses. The establishment of the grass species found in Malesia and China today depended upon the adventitious and recent introduction by man from the seed reservoirs of Africa and western Monsoon Asia, and perhaps to a lesser extent from monsoonal Australia. It is against this historical background that analyses of grass floras of south-east Asia have been made (9.5.2).

2.5 **Western Monsoon Asia**

This is the region (Bangla Desh, Burma, India, Nepal, Pakistan and Sri Lanka) which has been well covered as far as taxonomy is concerned, in BOR's classic volume (1960). The grass covers of India have also been studied in a reconnaissance survey, which has now clarified the true place in succession (5.8.3; 5.9.2 and Table 5/2) of the innumerable seral stages which are to be found in a country under great and increasing demographic and biotic pressure (DABADGHAO and SHANKARNARAYAN, 1972). In the present context, it is necessary to study the evolution of these major and secondary types of grass cover, using the same technique of ecological/historical analyses.

Speaking of the origin of Indian grasslands, BOR (1960, pp. 31–3) notes that species found in grassland in forest climates belong to the more highly developed genera.

'Their migration to the open country has tended to vest them with adaptations which are not necessary in . . . the tropical forest in which their ancestors presumably arose.'

Ecologists believe that the most advanced species taxonomically are found among communities low in the succession, particularly in arid and semiarid environments; there, physiological stress leads to chromosome abnormalities, which together with hybridization, provide the variability from which new speciation may evolve (12.8).

The Indian subcontinent has no grassland climaxes, being covered with many types of forêts denses, forêts claires and desert vegetation in which grasses may or may not play an important part as a ground storey. However, the areas at present covered by these forest types (CHAMPION & SETH, 1968) are almost certainly quite different from those that they occupied before the anthropogenic factors became so intense. Again, we need to classify the forest types, past, and present, of the subcontinent on the basis of light availability at the forest floor before deciding which shade-loving species of Gramineae might be or have been found in those forest types providing their light requirements. BOR (1960, p. 52) lists some twenty grass species which are found only 'in the deep shade of the forest' in the Indian subcontinent. The list contains species noted for the fringes of the tropical rain forest in Malaysia (9.5.2), of the genera *Centotheca*, *Leptaspis* and *Lophatherum*. BOR gives their global distribution as follows:

Centotheca lappacea	– widely distributed in south-east Asia, China, Polynesia and tropical Africa
Leptaspis cochleata	– Sri Lanka to tropical Africa and Madagascar
L. urceolata	– Sri Lanka and south-east Asia to Solomon Islands
Lophatherum gracile	– north-east Himalaya, Khasi-Naga Hills, Burma, Malaya, Sri Lanka, China, Japan, Australia
L. gracile var. *zeylanicum*	– endemic to Sri Lanka

The early history of the Gramineae in western Monsoon Asia has to be considered in relation to two major tectonic events. The first was the rafting of what is now peninsular India from the southern hemisphere across the Equator to join southern Asia. Any grasses of temperate or xerophytic, subtropical adaptation which may have existed must have been eliminated when the Indian plate passed through the tropics. During this period of movement, the plate was covered with tropical wet evergreen forests with slight admixture of deciduous but still grassless tree communities.

The second major event, the uplift of the Himalaya, caused progressive change in the insular Gondwana plate towards the modern, semi-arid monsoonal climate of the subcontinent (VISHNU-MITTRE, personal communication, 1973). As the Himalaya approached a certain height (probably about 4,000 m.), a rainshadow was created in the lands to the north, which has become progressively more intense with the continued rise of the range into modern times, at the rate of 12 m. per century. Tertiary orogeny of the Himalaya and its associated ranges was the catalyst which led to the gramineous explosion in the subcontinent, and to the origin of the annual cereals in the heartland of Asia to the morth.

With the consequent change from humid tropical towards semi-arid conditions south of the Himalaya, first in the north-west of the subcontinent and spreading progressively to the east and south-east into modern times, the densely canopied, grassless tropical wet evergreen forests have been replaced by more open types of forests (forêts claires), which favour the growth of light-requiring, more xerophytic grasses. This advancing aridity, associated with gradually rising temperatures which have become intensified during the post-pleistocene or Neothermal period, has created physiological conditions of stress favouring rapid speciation (12.8), and has led to the evolution of the rich gramineous flora of the present day (about 250 genera and 1,250 species, according to BOR, 1960).

Since his appearance on the scene, man has cut fuel, burned, cleared for cultivation and grazed his domestic livestock. The intensity of these activities has increased catastrophically during the past hundred or even fifty years, thus exacerbating the aridity and its effects on the physiology, and therefore the genetics, of the gramineous plants.

3 EVOLUTION OF GRASS COVERS: AFRICA AND AMERICA

3.1 **Africa south of the Sahara**

3.1.1 Palaeoclimatic Hypothesis

It is possible to trace the evolution of the vegetation of Africa further back in geobotanical history. The present discussion relates to the history and evolution of the grass covers as proposed mainly by Aubreville (1962). The concepts of Jacques-Félix (1962) with regard to the tribes, genera and species of the Gramineae are given in 9.2.1.

Aubreville (1962) has proposed a palaeoclimatic origin for certain savannas which occur today in Africa. Where soil is not a factor and the influence of man did not early acquire a significant intensity, it is necessary to suggest another explanations for the vast areas of savanna now found in climates suitable for dense humid or dry forest. Some believe that the Quaternary glaciations would have brought humid conditions and only a slight fall in temperature in Africa and Latin America. Aubreville, on the other hand, considers that a polar shift of about 12.5° latitude south would have brought dry and very cold conditions to west and central Africa. This change would have been sudden, causing a retreat southwards of the forest, while its place was taken by grassland. The forest moved to the southern tip of the continent. The Angola coast remained covered in savanna and steppe in spite of being in the Equatorial zone. This was due to the desertifying influence of the Benguela current. The east African climate became favourable for the expansion of tropical forests on mountains and the high plateau, and their expansion towards the south where they approached the region of the Cape. (A possible ecological origin of the grass components of these derived African savannas is given in 5.3).

With the return of the Pole northwards and the retreat of the glaciers of the last Ice Age, Africa south of the Sahara again acquired a forest climate (Aubreville, *op. cit.*). This created conditions for the return of the many forest types of the region from the favourable mountain and coastal bastions where they had persisted, just as today there remain in favoured localities in southern Africa certain vestiges of the Guineo-Congolese forest. Forest regeneration proceeds along banks of water-courses, extending in corridors, taking over hills, coastal plateaux on young soils, more humid soils, and eroded areas protected from fire. Tentacles join isolating savannas, which are then colonized from the forest and taken over. The forest has not yet reconquered all its old domain (AUBREVILLE, 1968). Progression in succession is always slower than regression; therefore this change back to forest was probably slower than the original change from forest to savanna; it has, of course, als obeen impeded or even prevented by the actions of man in his attempts to retain the economically valuable savanna cover.

If there had been an ancient indigenous grass flora in South Africa, it would probably have been eliminated at this time, its place to be taken later in the post-glaciation phase by grasses of east African origin.

When the forests returned to roughly their present position between Gabon and Uganda, between its southern borders and the limits of the position occupied during the Quaternary glaciation, a vast region was invaded by grasses. Then the semi-xerophytic flora of the present dry forests, open forests and Sudano-Zambesian wood savannas typified by *Brachystegia* also pushed north into the newly 'savannized' region. This progress is, however, less rapid than the withdrawal of the humid forest. That is why there persists between the southern borders of the humid forest and the forêts claires of *Brachystegia* characteristic of Angola, Katanga and Zambia, a great expanse of grassy or poorly wooded savannas – a biological hiatus. DUVIGNEAUD (1951) has noted that the Angolo-Rhodesian open forest is moving north across the great savannas, and has not yet established contact with the equatorial forest.

3.1.2 GRASS COVERS IN NATURAL VEGETATION

It is necessary to consider the ecological status of the grass covers of Africa south of the Sahara in the natural vegetation (RATTRAY, 1960). 'Dans les régions intertropicales le climax est de façon générale, constitué par des forêts. Le fait est unanimement admis par les botanistes et les forestiers' (SCHNELL, 1970/1971).

AUBREVILLE (1962) enumerates the climax tropical vegetation types as follows:
'Les climax forestiers dans la zone intertropicale
Nous arrivons alors à ces conclusions générales très importantes d'écologie tropicale que dans les pays tropicaux la végétation primaire est sous les climats humides une végétation de forêt dense, ou plus généralement de végétation ligneuse dense et, sous les climats semi-arides, encore une végétation ligneuse, biologiquement différente des précédentes, fermée ou ouverte; lorsqu'elle est ouverte il y a transition vers des formes mixtes graminéennes et forestières. Dans les régions semi-désertiques, les formations climax sont encore des steppes à végétation ligneuse, des sous-arbrisseaux nains et des succulents, très ouvertes, où les herbacées pérennes (graminées) n'ont plus une place préponderante, alors qu'au con-

traire les herbacées annuelles abondent en saison estivale pluvieuse. Tels sont à notre avis les grands types climaciques de végétation tropicalé.'

Wooded savanna and open forests are climax only in very arid country, where due to poor soil moisture, the roots of shrubs and trees spread, leaving gaps between them which are occupied by grass. Elsewhere a mixture of shrub/grass is an anomaly. The forest types will eliminate the grass, in the absence of repeated bush fires and clearing. In actual fact, however, the forest rarely takes over savanna, and thus a fire subclimax acquires an apparent stability in the form of wooded savanna. It has been shown that protection from fire in Ivory Coast, Katanga, Zambia, Zaire and Nigeria leads to the regeneration of tree cover (AUBREVILLE, 1966a). ADEJUWON (1971) has suggested that the climax vegetation in the savanna zones of Nigeria (Sudan, Sahel, north and south Guinea) was tropical, xerophytic woodland, tropical deciduous woodland and tropical rain forest respectively.

In their survey of the land resources of Lesotho, BAWDEN & CARROLL (1968) state that this is a grassland country and there is an almost complete absence of natural tree growth. ACOCKS (1953) has suggested that conditions are too dry and frosty for tree growth and that most of

Fig. 3/1 *Key**

Desert and sub-desert climates 5_1 5_2 4_1 4_2

Desert climate: annual precipitation below 200 mm. 11 to 12 dry months. Some years without rainfall. 5_1 5_2

Sub-desert climate: annual precipitation between 200 to 400 mm. 9 to 11 dry months. 4_1 4_2

Dry tropical climates

Usually two very rainy months, receiving most of annual precipitation of between 400–1200 mm.
4 to 5 dry months: $3a_1$ $3a_2$
6 to 8 dry months: $3b_{11}$, $3b_{121}$, $3b_{122}$ $3b_2$

Greater variability in no. of dry and wet months: $3c_1$ $3c_2$

Semi-humid tropical climates

Precipitation from 900 to 1400 mm. 4 to 5 dry months.
humid tropical, with cool season due to mountain massifs: $1a_{21}$ $1a_{22}$
subtemperate: 2_3

Affected by altitude 2_{22}, 2_{23}

* For details of seasonal variation in temperature, see original text.

Fig. 3/1: Map of the intertropical climates of Africa (BILLE, LEBRUN & RIVIÈRE, 1968/69, after AUBREVILLE, Climats, forêts et désertification de l'Afrique tropicale)

Table 3/1

Woody types of vegetation in difficult environments in Africa (AUBREVILLE, 1962)

	Climate	Rainfall mm.	Duration of ecologically dry season months	Annual saturation deficit and regime of monthly deficit
Wooded savannas and Sudanian open forests	Sudano-Guinean	medium or high. 950–1750	4–5	high: low or medium in rainy season, very high in dry
Open forest and wooded savanna of Upper Katanga and Zambia	Upper Katangan	medium 1000–1400	5–6	medium: high at end of dry
– id – Zambia and Rhodesia	Rhodesian	low or medium 600–1200	6–7	medium: high at end of dry
Steppe. Acacia and Sahélo-Sudanian wooded savanna	Sahélo-Sudanian	400–1000	6–7	very high, sometimes excessive
Thickets. South Madagascar, Didieraceae and Euphorbeae	South Madagascar	500–550	6–7	medium

the grasslands of Lesotho are therefore 'climax' types. Other southern African botanists have preferred to regard them as a fire sub-climax grassland (RATTRAY, 1960). A sub-climax succession to shrubs may occur, in sufficiently sheltered spots at any altitude if fire is excluded. Many years of overpopulation, over-grazing and veld fires have greatly influenced the development of the present vegetative cover and most of the shrubs that now occur are unpalatable to stock. AUBREVILLE (1962) notes that in the east, rainfall 730 to 1,500 mm. and dry season of four to five months, the climax should be dry forest or at least open forest. More to the west, with rainfalls down to 450 mm., with fairly high saturation deficit and coldest average month 4 to 7°C, there are a number of mountain tropical woody formations that are adapted.

WALTER (1969) claims that in South-West Africa climate is the dominant factor governing the distribution of grass and shrub communities. At 100 mm. annual rainfall, the grasses take up all the available soil moisture. At 200 mm. the same situation occurs, but with taller grasses. Only at 300 mm. is sufficient water available to permit small woody species to enter the communities. Shrubs become taller with increase in rainfall, until tree savanna is reached. In all cases, grasses are dominant, with non-woody plants using the residual moisture (WALTER, *op. cit.*). Between 500 and 600 mm. rainfall, trees develop until their canopies touch, in a type of deciduous tropical forest from which grasses are eliminated by shading. AUBREVILLE (1962) considers that the south-west African coast and the Karroo remained desertic during the palaeoclimatic shift of the Quaternary, under the influence of the cold Benguela current. This explains why today the biological spectrum of the vegetation is so different from the Sahara, which is a young desert having undergone frequent changes since the Tertiary.

3.1.3 EDAPHIC CLIMAX GRASS COVERS

Many writers refer to the existence of edaphic climax types of grass communities, more especially in Latin America (3.2.2). AUBREVILLE (1962) stated that, in general, edaphic savannas cover only small areas and appear as clearings in forest formations. Some soils are indeed not adapted to woody vegetation, e.g. where an impermeable layer is near or on the surface; the dembos in slight topographical depressions, warped, where rainwater stagnates; shallow sandy soils covering a shallow clay layer, poorly drained, very dry during the dry season and saturated or even flooded in the rainy season. Sometimes an impermeable layer surfaces as a non-fissured ferruginous plate. Only grasses can live in small cracks in the surface where a little soil has accumulated. Sandbanks which have recently re-emerged from water have grasses as the pioneer vegetation.

3.1.4 THE FOREST/SAVANNA BORDER

Throughout the intertropical region, one is impressed by the sharp break which occurs between dense tropical forest and savanna; the absence of intermediate forms of forest or shrubby communities appears as an ecological anomaly (HILLS & RANDALL, 1968).
AUBREVILLE (1966a) has considered this problem with special reference to West Central Africa, the Guineo-Congolese vegetation zone. Here the climatic gradient is gradual without sharp changes in ecoclimate which could account for this phenomenon. Quite distinct soil types occur over small areas, but these are found in both forest and savanna. A phytogeographical hiatus is found between the dense humid tropical forest on the Congo-Guinea zone and the open forest and woody savanna of the Sudano-Zambezian (see LETOUZEY, 1968, for a phytogeographical study of these types in Cameroon). Even in its arid interior, the Sudano-Zambezian zone has a relatively rich flora, with genera and species which may be endemic although related to the Guineo-Congolese flora. Between these two chorological regions there exists a gap,which is occupied by sparsely wooded savanna with a limited number of species.

Fire has in most areas eliminated the transition from dense humid to dense dry forest, and all stages in the succession to the semi-arid, subdesertic tree savanna. Annual fires generally prevent colonization from fire-sensitive forest pioneers, while the dense humid forest forms an impassable barrier to savanna vegetation. Hence the comparative floristic poverty of the savanna border (AUBREVILLE, 1966a).

3.1.5 ANTHROPOGENIC FACTORS

The outstanding factor in the evolution and maintenance of the grass cover of tropical Africa is, of course, fire. Anthropologists report the practice of burning vegetation among the customs of primitive communities from the earliest times. CLARK (1970) states that man has used fire for well over 50,000 years, while during Neolithic times it was a common tool in land management in north and sub-Saharan Africa. However, dense humid tropical forest could not have been destroyed by fire until primitive peoples possessed iron tools for felling, and these must have come from arid regions (WHYTE, 1972b).

The plant geographers recognize that fire has played a predominant role in the creation of the vegetation of Africa as it occurs today:

'Il parait aujourd'hui indiscutable que la végétation climatique, c'est-à-dire le manteau végétal naturel, est en général forestier. Les savanes, les prairies dans l'ensemble, paraissent bien n'être que des paysages secondaires, issus d'une déforestation et entretenus dans leur état actuel par l'action directe ou indirecte de l'homme' (SCHNELL, 1970/1971).

The managers of grazing lands have adopted fire for the maintenance of a grass cover in a vegetation which is always struggling to revert upwards in succession to a forest climax, through scrub, shrub, thicket, thorn forest, etc. Fire has been used for several purposes:

a. to concentrate wild animals for easier slaughter by hunters;
b. to control ligneous growth and so reduce competition;
c. to promote regeneration of grass growth in tall grass savanna, and
d. to clear and fertilize land for shifting cultivation.

(see also 6.5).

SCOTT (1970) quotes earlier writers as saying that veld fires have been a feature of the African landscape since time immemorial. Early Portuguese explorers saw that the interior of South Africa was covered by a pall of smoke due to veld burning, and called the country Terra dos Fumos. RAMSAY & ROSE-INNES (1963) have reviewed the literature on grass and bush burning in West Africa. They describe the principal types of subclimax and climatic climax vegetation in the northern Guinea savanna zone of northern Ghana, and discuss the physical and biotic factors which have influenced this. Quantitative analyses have been made of the effect on degraded Guinea savanna vegetation of annual early dry-season fires, annual late dry-season fires, and complete protection. ADEJUWON (1970) states that the coastal savannas of Nigeria are not the result of human activities, but then admits that they are maintained as a subclimax by annual fires.

LEMON (1968a, b) has considered fire on the Nyika plateau in Malawi in relation to the pattern of wildlife grazing. Fire is an important agency of control, if the objective is to prevent animals from wandering off the central grasslands during the dry season into areas where they will certainly be subject to poaching.

LAMOTTE (1967) states that fires are prevented to promote comparative studies of unburnt and burnt grasslands in an area around the Lamto Ecological Field Station in Ivory Coast. Annual rainfall varies from 1078 to 1657 mm. with two dry seasons, a long one from November to March and a short one in August. Gallery forests are found along the Bandama River and its tributaries; elsewhere the open savanna is interspersed with numerous *Borassus* palms.

There are two possible types of savanna:

1. those forming in old, dry forests, including species partially resistant to fire; more densely wooded and floristically rich; and

2. those appearing after cultivation on the borders of the humid forest; derived savannas later colonized by ubiquitous species from drier regions; floristically poor.

With time the differences may become less marked, since a dry forest flora has a tendency to spread into all savannas, in spite of fire (AUBREVILLE, 1966a):

Where human activities (cultivation and burning) are absent, forest regeneration is rapid, e.g. in protected enclosures in the Congo, Gabon, the High Sangha, North Cameroon and Uganda. In the process of regeneration, the trees of the savanna (e.g. *Terminalia* in Uganda) are completely eliminated as the true forest vegetation advances.

AUBREVILLE (1966a) has summarized his conclusions on the savannas of Africa: 'Sauf le cas de savanes herbeuses édaphiques, les savanes herbeuses ou peu boisées, ayant une flore forestière pauvre ne sont pas des formations climaciques. Elles ont une origine anthropique contemporaine ou peut-être une origine paléoclimatique récente, disons du Quaternaire le plus prôche'.

3.2 America: South, Central and Caribbean

3.2.1 SAVANNA – DEFINITIONS

The criteria applied in the definition of savanna in tropical South and Central America are partly geographic and partly derived from the physiognomy and structure of the vegetation (VAN DONSELAAR, 1965); the only point of agreement is that there is a ground layer, and that this ground layer is always ecologically dominant. These definitions have been made by botanists and phytogeographers; geographers such as JAEGER (1945), TROLL (1952) and LAUER (1952) have adapted the nomenclature to their specialized disciplines, in particular regarding the relation between vegetation and climate.

(i) LANJOUW (1936) 'savannahs are plains in the West Indian Islands and northern South America covered with more or less xeromorph herbs and small shrubs and with few trees or larger shrubs.'

(ii) DANSEREAU (1951) 'savanna: more or less closed ground layer, very discontinuous upper (woody) layer.'

(iii) BEARD (1953) 'savannas are communities in tropical America comprising a virtually continuous, ecologically dominant stratum of more or less xeromorphic plants, of which grasses and sedges are the principal components and with scattered shrubs, trees or palms sometimes present'.

(iv) EECKHOUT (1954) 'savanna: open xerophytic herb formation in the tropics with scattered xero- or tropophytic trees or shrubs, isolated or in clumps.'

(v) DYKSTERHUIS (1957) 'xeromorphic grassland containing isolated trees.'

(vi) SILLANS (1958) 'formation characterized either by a continuous herb layer, mainly consisting of more or less high and more or less dense grasses, or in addition by a layer of shrubs or by a layer of trees of very variable density.'

(vii) Fosberg (1958) 'closed grass or other herbaceous vegetation with scattered trees.'

(viii) Heyligers (1963) 'tropical vegetation that is neither swamp nor primary nor secondary forest.'

(ix) van Donselaar (1965) 'a savanna or a campo is an area with a xeromorphic vegetation comprising an ecologically dominant ground layer consisting mainly of grasses, sometimes together with sedges, and with or without trees and/or shrubs either forming a more or less continuous layer, or occurring in groups, or isolated.'

(x) Walter (1969) 'tropical vegetation community, ecologically homogeneous, grass component dominant and woody plants dispersed among Gramineae, all growing in identical environmental conditions; savannas contain two types of plants in labile balance competing for water; to understand their interrelations, the ecophysiology of the Gramineae and the typical woody plants must be known.'

3.2.2 Origins of savannas

It is appropriate to give the impressions of a specialist on African savannas after studying the savannas of tropical America, and to compare these with a presentation of a specialist in South American savannas (Blydenstein, 1968), made on the basis of local experience and a study of African literature.

Aubreville (1962, 1965) believes that, where savannas are found in areas of very differing soil and relief, with a forest climax climate, where the forest has disappeared and where population is low, the same palaeoclimatic origins as those noted in Africa apply. With the Quaternary polar shift and the movement south of the Equator some 8° in Latin America, the pine forests of south Brazil were eliminated and their place taken by savannas. In Central America (Costa Rica, Nicaragua and Honduras), pine savannas occur in an area with a wet tropical climate which should maintain tropical rain forest. Aubreville (1966b) states that the indigenous peoples of these areas could not have burnt the tropical rain forest. Therefore the reason for the occurrence of a pine forest in a tropical rain forest climate is palaeoclimatic in the first instance, perpetuated by fires to produce pine savannas over the subsequent millenia (Taylor, 1962).

The Sabana Grande took the place of humid tropical forest in the Guyanas. The northern coastal areas within the intertropical zone still received abundant rainfall, and here the dense tropical forest remained. Amazonian Hylaea moved southwards, joining the Atlantic forest of the east and south, which persisted under maritime influence. The zone between the Guyana forest and Hylaea, marked today by predominantly xerophytic vegetation (Llanos, the savannas of Ampapa, caatingas of Piauhy and Ceara, the pseudo-caatingas of Upper Rio Negro, the campos of Upper Rio Branco, the campos and dry forests of the middle Amazon) was probably drier at that time, while the basin of the middle Amazon was more or less deforested.

Faced today with clear evidence of the progressive destruction by fire of the Venezuelan forests and the extreme poverty of the ligneous flora of the Llanos, AUBREVILLE (1965) considers it inconceivable that the savannas are climax, as many phytogeographers still believe.

'The tropical wet and dry climate, with its alternating rainy and dry seasons, is a prerequisite for good burning conditions . . .. The typical wet and dry climate has been called a 'savanna climate' because the long dry season following a continuously rainy wet season was considered 'hostile to woodland' (SCHIMPER, 1903–1904). In actual fact, the major part of the region with a tropical wet and dry climate is covered by deciduous forest, or at least its visible remains . . .. Although there is little argument about the dominant role of fire in the environment of savannas, once these are formed, the role of fire in the destruction of forests must be studied in the forests which are supposedly destroyed . . . trained observers have been able to recognize the signs of forest relicts within these grasslands and the general tendency to vegetation succession in the direction of forest . . . such degraded forest is common, and is usually associated with cultivation and forest clearing practices, among which fire plays an important, but not exclusive role' (BLYDENSTEIN, 1968).

The floods of the western Llanos discourage forest; higher areas support forests, however, and dense semi-deciduous forests exist in the western plains, even in more or less flooded areas.

The origin of the Llanos is frequently attributed to edaphic causes, due to the great extent of the impermeable layer or hardpan, and a succession of alternating drought and flood which are inimical to tree growth. In Surinam, TEUNISSEN & WILDSHUT (n.d.) state that the hydrology of the soil, as governed by the presence or absence of a more or less impermeable soil layer at shallow depth, is a principal habitat factor influencing the location of xerophytic forest, savanna scrub, savanna bushes and open vegetation, and exclusively open vegetation.

VAN DONSELAAR (1965, 1968, 1969) states that the climate of northern Surinam is a 'savanna climate' characterized by a certain difference between the precipitation in dry and wet seasons, independent of absolute values. A savanna vegetation is natural, i.e. determined edaphically, if the upper layer of the soil is alternately desiccated and saturated, in wet and very wet localities. A savanna vegetation occurs in dry localities only if fires prevent the formation of a closed layer of shrubs or trees. The origin of savanna formations on white sands in Surinam is discussed by HEYLIGERS (1963)(see 5.8.5 for grass communities).

The presence of trees has been attributed to their possible ability to penetrate the hardpan (WALTER, 1969). BLYDENSTEIN (1962) has noted that the watertable at the beginning of the rains is–575 cm. and at the end–385 cm., and that the hardpan cannot therefore be entirely impermeable. However, during trials at the Estación Biológica de Los Llanos near Calabozo, Venezuela, it was discovered that tree and isolated bush roots do not penetrate the layer, while they may travel laterally as far as 15 m. from a bush only 6 m. high. Moreover AUBREVILLE (1965) has observed dense dry forest or at least thickly wooded savannas and open forests in Africa, India and Cambodia on laterite shields with very thin soils, in climates drier than that of the Llanos. Copses may be seen in the Llanos during the dry season which are still green, showing that adequate soil humidity is available. Here again, fire appears to be the dominant ecological factor determining the savannas of the Llanos.

As in Africa, the poverty of the floristics of the Llanos presents an anomaly, contrasting sharply with the rich, ancient, autochthonous flora of arid and semi-arid South America – for example, the caatingas and the campos cerrados of Brazil (600 species belonging to 336 genera and 83 families – GOODLAND, 1970). The shrubs, *Curatella americana*, *Bowdichia virgilioides*, *Byrsonima* and *Roupala* occur in all poorly shrubbed savannas of French Guyana, Ampapa, Upper Rio Branco and in the Llanos; they are recent colonizers, adapted to the now mediocre soil conditions of the deforested lands. There are indications that the germination of *Curatella* seeds is improved after passing them through a flame (BLYDENSTEIN, 1968).

The presence of dry deciduous forest in Guarico and Anzoategui is taken as confirmation of the fact that, several thousand or tens of thousands of years ago, the Llanos was occupied by a dry forest formation of the type which remains today in thickets, grading towards the east into more humid forest at the foot of the Andean Cordillera, in the Orinoco delta and in Guyana (AUBREVILLE, 1965).

The origin and present status of the Rupununi savannas in former British Guiana (now Guyana) have been studied by FANSHAWE (1952), and later by the Regional Research Centre of the British Caribbean at the Imperial College of Tropical Agriculture, Trinidad (LOXTON, RUTHERFORD & SPECTOR, 1958; STARK, RUTHERFORD, SPECTOR & JONES, 1959). Particulars are given of climate, geology, geomorphology and soils. The vegetation of the northern, southern and central savannas is similar; treeless stretches of grassland are common: *Trachypogon plumosus* dominant, with *Aristida setifolia*, *Axonopus aureus*, *A. compressus*, *Andropogon angustatus*, *Mesosetum loliiforme*; on higher, better-drained soils the fire-resistant tree, *Curatella americana*, in open formation, with occasional *Byrsonima crassifolia*; *Anacardium occidentale* common around derelict house sites, ground cover of *Axonopus compressus*. The Good Hope –

Annai Mountains carry what is considered by MYERS (1936) to be a deciduous monsoonal type of forest, similar in composition to fringing forests along rivers – great mixture of shrubs, small trees and lianas, with many Myrtaceae and Leguminosae.

EDEN (1970) states that climatic fluctuations during the Quaternary and resultant changes of soil water conditions must be considered as possible causes of vegetational change in the Rupununi. Pollen data suggest that savanna or dry *Byrsonima* forest has existed in the northern Rupununi from late glacial times to the present, possibly pre-dated by a more extensive, semi-deciduous forest now present as relic 'bush islands'. One has the impression that EDEN does not give enough discredit to adverse soils and the much more recent burning which have contributed to the establishment and maintenance of the present savanna formation.

The savannas of Venezuela have been described by RAMIA (1961, 1964, 1968), BLYDENSTEIN (1961, 1962, 1963; see 5.8.5 for subtypes of *Trachypogon* savanna) and GOODLAND (1966). The Llanos Orientales of Colombia (FAO/UNSF 1966; BLYDENSTEIN, 1967) contain, in addition to forests, savannas in which annual grasses are dominant:

savannas with residue of forests
- (i) *Melinis minutiflora*
- (ii) *Paspalum carinatum*
- (iii) *Trachypogon ligularis/Paspalum carinatum*

savannas subject to flooding
- (iv) *Andropogon*
- (v) *Mesosetum*

humid savannas
- (vi) *Leptocoryphium lanatum*
- (vii) *Trachypogon ligularis*
- (viii) *Trachypogon vestitus/Axonopus purpusii*
- (ix) *Paspalum pectinatum*
- (x) *Trachypogon vestitus*

A great number of species are shared by the last seven types.

BLYDENSTEIN (1967) discusses theories on the origin of tropical savannas in relation to the environment (climate, landscape, including geological history and geomorphology, soil, fire and man). Pollen analyses have shown a dominance of Cyperaceae and Gramineae in an herbaceous vegetation to the northern edge of the Llanos in Pleistocene times. COLE (1960) postulates that grass savannas in Brazil represent the older vegetation formations, which are now gradually being replaced by forest on dissected terrain as the level peneplain becomes eroded. It is now considered that man arrived on the American continent between 40,000 and 15,000 years ago, and that the palaeo-Indian epoch in the Llanos was from 15,000 to 5,000 B.C. (BLYDENSTEIN, 1967).

A modern comparison has been made in Costa Rica between environments of a tropical semi-deciduous forest with a contiguous area of derived *Hyparrhenia rufa* savanna which has been burned annually (DAUBENMIRE, 1972b). During the five-month dry season the savanna soil becomes hard and cracks. Compaction by zebu cattle and horses, coupled with a very sparse, invertebrate soil fauna, reduces porosity. During the season of maximum leaflessness, the forest still intercepts about 33 per cent of solar radiation; the savanna soil is rendered bare and black by fire during this season. Higher soil temperatures and the annual burning probably account for much of the striking reduction in burns in the savanna soil profile. Soils become dried to wilting point to a depth of over 90 cm. in the savanna, but remain moist below 30 cm. in the forest. No downward movement of clay or other irreversible changes in the profile have been noted. There is, however, good evidence of a loss of 11 cm. of soil during the 22 years since deforestation; this is regarded as the most significant aspect of environmental deterioration.

The most characteristic herbaceous plants of the savannas in Cuba (VORONOV, 1970) are grasses of the genera *Panicum*, *Paspalum*, *Arundinella* and *Arthrostylidium*. On the poorer soils, palm savannas (genus *Sabal*) occur, possibly originating from palm groves due to anthropogenic influences. Relations are not quite clear between the savannas and the thorny shrub communities, which occur mainly in tropical conditions and generally indicate increasing aridity. Savannas with low or stemless palms (*Copernicia* spp. with a herb/ low shrub layer) are formed on serpentines, a markedly deviant, edaphically conditioned group of formations rich in endemic species. Gradual salinization of the soil in savanna swamps, usually stagnant, leads to the formation of solonchak meadows; grasses with rigid stalks and short leaves – *Distichlis spicata*, *Sporobolus domingensis*, *Cynodon dactylon* and *Paspalum vaginatum*.

3.2.3 CERRADO OF BRAZIL

The vegetation of almost all Central Brazil is cerrado (meaning in Portuguese half-closed, dense or ajar), an appropriate name for this vegetation which is neither open nor closed, that contains every physiognomic type between these two extremes, herbaceous, grassy, shrubby, orchard, woodland, almost to forest (GROSSMAN, ARONOVICH & CAMPELLO, 1966; LIMA, 1966; SANTIAGO, 1970; GOODLAND, 1970, 1971a and 1971b).

EITEN (1972) argues that the cerrado is a climax vegetation, not a stage or a disclimax in succession to forest or any other vegetation type. Both in its central position in the country and on its margins when it is in contact with other vegetation provinces, it is an original, primitive vegetation – a whole floristic province of semi-deciduous, xeromorphic vegetation. It has not arisen following the destruction of mesophytic forests by burning.

GOODLAND has studied in particular the oxisoils and the plant species and vegetation types of the Triângulo Mineiro. The physiognomic gradient parallels a soil fertility gradient; basal area per hectare is correlated with P, N and K (GOODLAND & POLLARD, 1973). Soil fertility is not necessarily the cause of the physiognomic gradient; differences in physiognomy and fertility may have arisen because of different histories of burning and cutting. The cerrado situation seems to contradict the statement of BEARD (1953): 'Savannas occur in ill-drained country ... the chemical status (of the soils) is of little account.'

In the cerrados, several species of grass may reach 3 m. in height: *Tristachya leiostachya*, *Axonopus compressus*, *Tristachya chrysothrix*, *Hyparrhenia rufa*, *Andropogon lateralis*, *A. condensatus* and *Melinis minutiflora*, in order of decreasing height. Grasses are reduced in number and height in other types, probably due to shading. The fact that woody vegetation is dominant in most Brazilian cerrados may distinguish them from the floristically related vegetation of the Llanos of Venezuela and Colombia discussed above. The rich flora of this area in Brazil would suggest great venerability, following AUBREVILLE. See bibliography no. 1268 of the Commonwealth Bureau of Pastures and Field Crops for 116 references on the llanos of Colombia and Venezuela and the cerrados of Brazil (1939–70).

3.2.4 TRANSITIONAL GRASS COVERS

The natural region called 'Gran Chaco' occupies large areas in Bolivia, Paraguay and Argentina. MORELLO (1968) has divided the Argentine Chaco into 'great units of vegetation and environment,' considering vegetation 'as the most sensitive and simplest indicator for registering and integrating environmental variations.' The model concept (roughly a group of recurrent patterns of vegetation) and 'three levels of perception' have been developed. The names of the grass genera of the Argentine Chaco would suggest at least a borderline subtropical environment: *Aristida adscensionis*, *Elionurus adustus*, *E. muticus*, *Panicum prionitis*, *Paspalum intermedium*, *Setaria* and *Sorghastrum* (MARLANGE, 1971). The general state of specific degradation and low nutritive value of pastures in Argentina has been confirmed by enclosures (BRAGADÍN, 1959) for the region of Los Llanos in the province of La Rioja. The species of Gramineae noted in the various closures were: *Aristida adscensionis*, *A. medocina*, *Bouteloua aristoides*, *Cottea pappophoroides*, *Chloris polydactyla*, *Digitaria californica*, *Diplachne dubia*, *Eragrostis cilianensis*, *E. longipila*, *E. virescens*, *Gouinia paraguayensis*, *Pappophorum* (= *En-*

neapogon) spp., *Paspalum unispicatum*, *Setaria argentina*, *S. geniculata*, *Sporobolus pyramidalis*, *Tragus racemosus*, *Trichloris crinita*, *T. pluriflora* and *Tripogon spicatus*.

It is in the Chaco that human and geomorphological disturbances have led to the creation of very unstable and non-adjusted marginal ecosystems with regard to structure and niches. MORELLO & SARAVIA TOLEDO (1959) have examined the changes caused by extensive livestock husbandry in the woods and grasslands of the Chaco. The primitive woods were of two types:

a. Quebracho stands of *Schinopsis quebracho-colorado* and *Aspidosperma quebracho-blanco*, and
b. Palosanto stands of *Bulnesia sarmienti* with either *Aspidosperma quebracho-blanco* or *A. triternatum*.

The primitive grasslands were composed of *Pennisetum frutescens*, *Elionurus tripsacoides*, *Trichloris crinita*, *T. pluriflora*, *Setaria argentina*, *Gouinia paraguayensis* and *G. latifolia*. In the upland savannas there is always evidence of fire; they are therefore considered to be pyrogenic in origin, and caused by lightning rather than man.

In the first quarter of this century, animal husbandry was limited to the grasslands, which were transformed into shrub communities of *Acacia*, *Mimosa*, *Mimozyganthus* and *Celtis*. Where these pastures were exhausted, the cattle began to feed in the interior of the wooded areas, inducing many floristic and plant sociological changes, e.g. propagation of species hitherto unknown in virgin forest, invasion by species largely restricted to the major drainage bottoms, invasion of arborescent Cactaceae, and the formation of a number of anthropogenic communities. MORELLO & SARAVIA TOLEDO (1959) continued their study by recording the present seasonal diets and grazing patterns of domestic animals in the forest communities of the Chaco, and the role of rodents as a destructive element in the peri-domestic areas, where one of the ten plant communities recognized is characterized by grasses of ephemeral type.

It is into some of the primitive systems that the woody colonizing species, *Prosopis ruscifolia* or vinal, has entered (MORELLO, CRUDELI & SARACENO, 1971). MORELLO (1970) has evolved a conceptual model of the relations between grassland and vinal (Figs. 3/2-5). This model considers that, before the pressure of cattle on the Chaco, the grasslands were regularly submitted to two types of periodic pulsations – fires and floods. Their disappearance would have started 'the explosive evolution of the woody plants in environments of grassland and tall grassland.'

Plant formations of potential value in Paraguay include the typical grasslands, the marsh vegetation and the palm savannas of *Copernicia australis*, *Acrocoma totai* and *Butea yatay*, the thorn savannas with various species of *Prosopis*, the tree savannas of *Tecoma argentea*, and the mosaic savanna of the 'montes', tropical, semi-deciduous, semiarid steppes (Fretes, Samudio & Gay, 1970). The Gramineae are present in greater number of genera, species (128) and individuals and occupy a much greater area than the herbage legumes (33 species); the latter are present solely as 'interstitial' components and have little effect on the quantity and quality of the forage (limited by acid soils, deficiency of phosphorus, calcium and possibly some essential minor elements). Woody species of *Prosopis*, *Mimosa* and *Acacia* are invaders in some alluvial

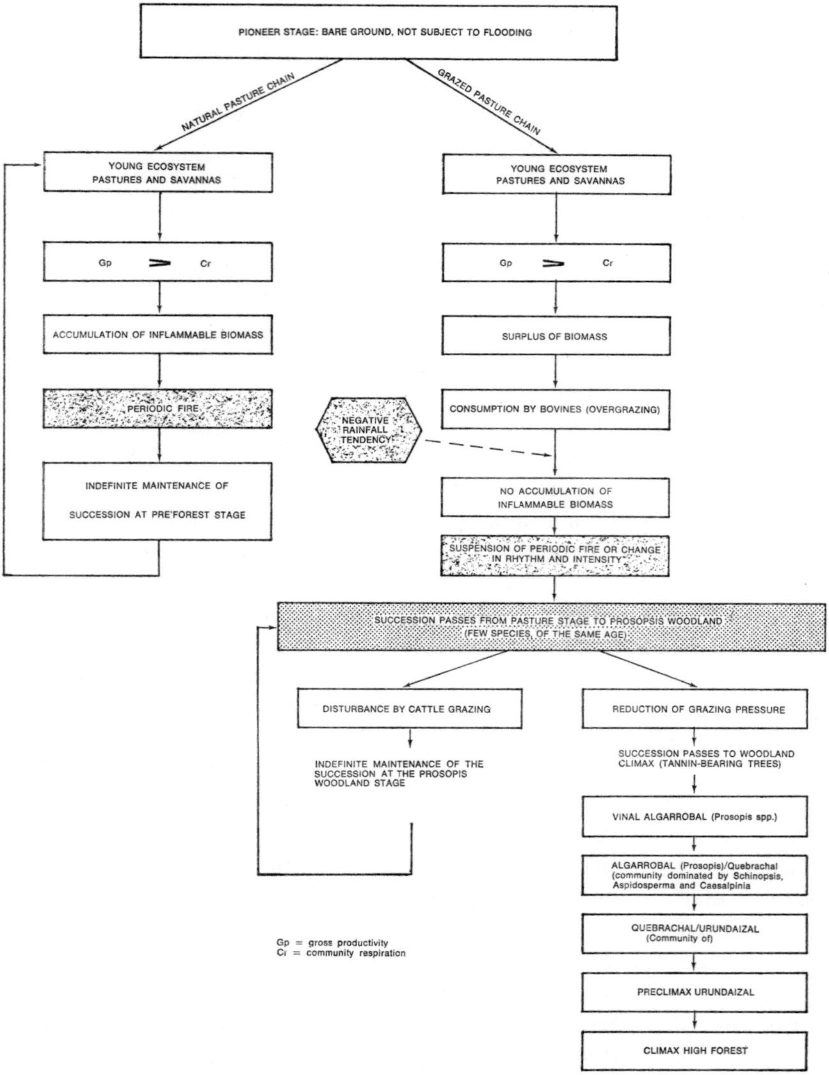

Fig. 3/2: Model for the evolution of an ecosystem on land not subject to flooding, with and without cattle (Morello, 1970).

grasslands. The genera of the Gramineae that are predominant in respect of area are *Paspalum*, *Andropogon*, *Sorghastrum*, *Elionurus* and *Axonopus*; in respect of number of species, *Andropogon*, *Eragrostis*, *Panicum*, *Paspalum*, *Aristida*. The nutritive value of grass species expressed in terms of crude protein at different seasons is shown in Fig. 3/6.

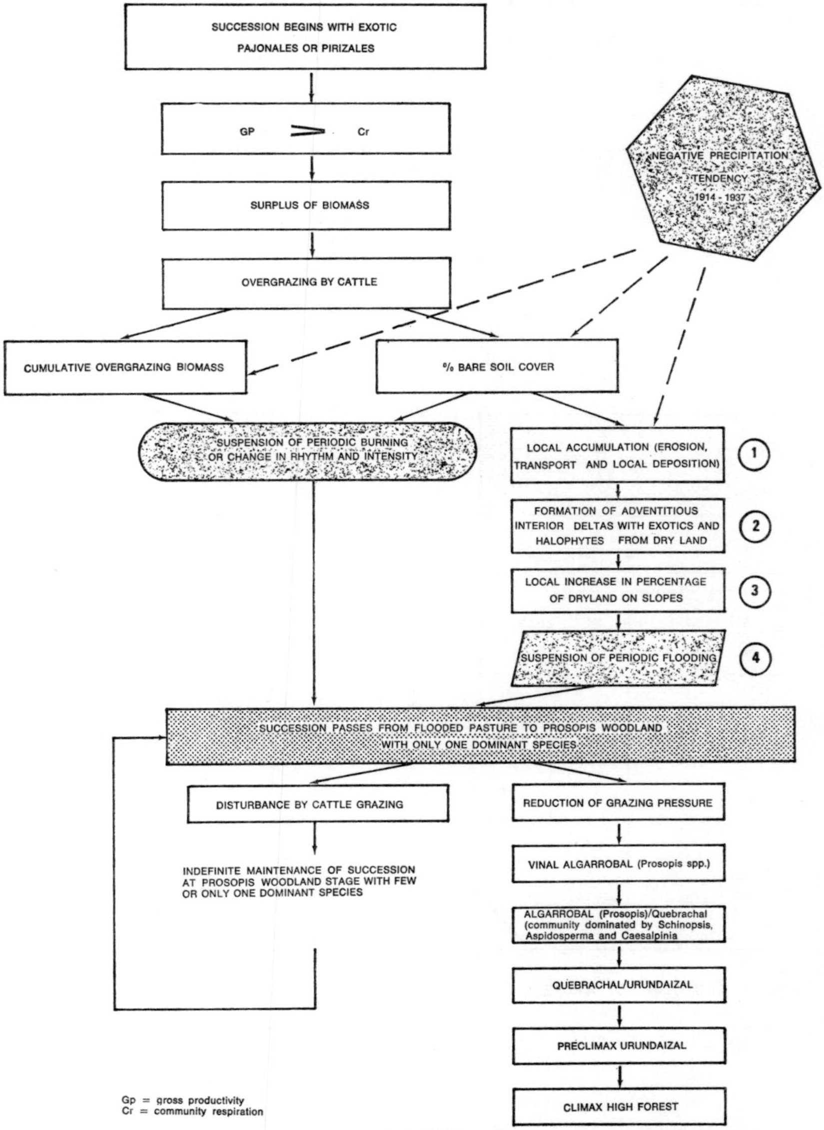

Fig. 3/3: Model for the evolution of an ecosystem on land subject to flooding and with overgrazing (MORELLO, 1970).

Walter (1967a and 1967b) considers that: 'the pampa is a transitional zone between the temperate grasslands caused by aridity and the tropical grasslands caused by topographical factors. The grasses can withstand the change from humidity to aridity during a year, but not the trees'.

But, according to Troll (1968b) all months of the year have a precip-

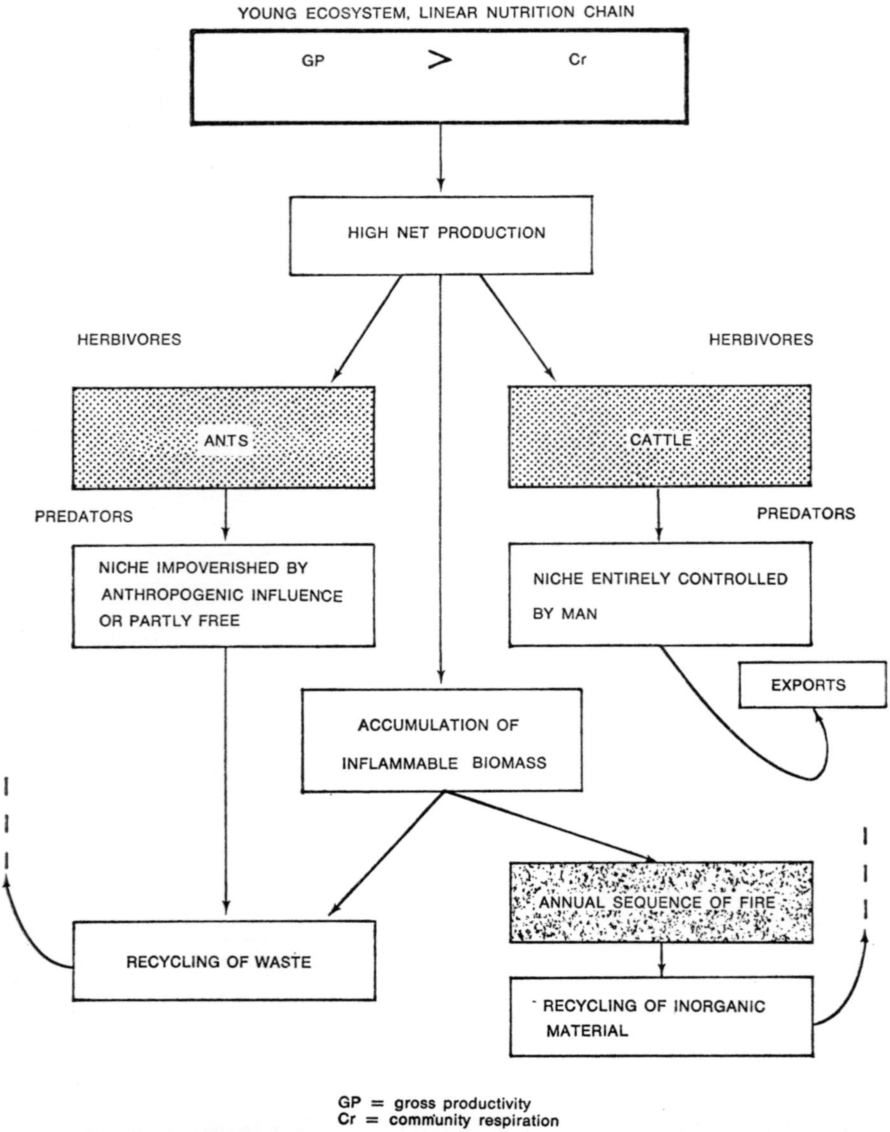

Fig. 3/4: Simplified model of present grazing systems in the Chaco pastures not subject to flooding (Morello, 1970).

itation far in excess of potential evaporation. ELLENBERG (1962) concludes that many native and exotic tree species would spread spontaneously 'if the omnipresent grazing cattle did not destroy the tree seedlings.' HUECK (1953) reaches similar conclusions regarding the Chaco and parklands of Tucumán Province; only human influence could have created the park landscape from former dense forest.

In her review of literature on the grasslands of Latin America, ROSEVEARE (1948) discussed the work of ROSENGURTT and his colleagues (1939; 1944) on the grass covers of Uruguay; this country is also borderline in the present context, since the grass covers are admixtures of tropical and subtropical (summer-grazing) and temperate (winter) species. The Mesopotamia region of Argentina has a transitional climate (VAN DER SLUIJS, 1971) towards a cool, steppe climate in the south and a cool, tropical climate in the north.

Transitional types and zones of vegetation in Mexico are described by LEOPOLD (1950), and HERNANDEZ EFRAIN (1963). BEARD (1944) considers that the study of communities in tropical America may offer the key to the understanding of African types of vegetation, particularly where fire and grazing are operative. PENNINGTON & SARUKHAN (1968) describe the various biomes in the Mexican tropics.

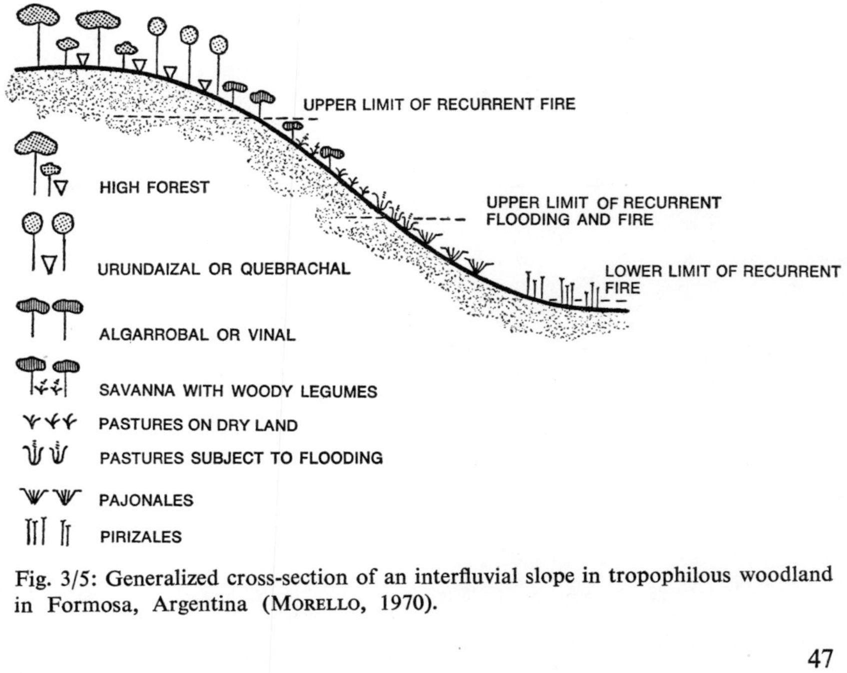

Fig. 3/5: Generalized cross-section of an interfluvial slope in tropophilous woodland in Formosa, Argentina (MORELLO, 1970).

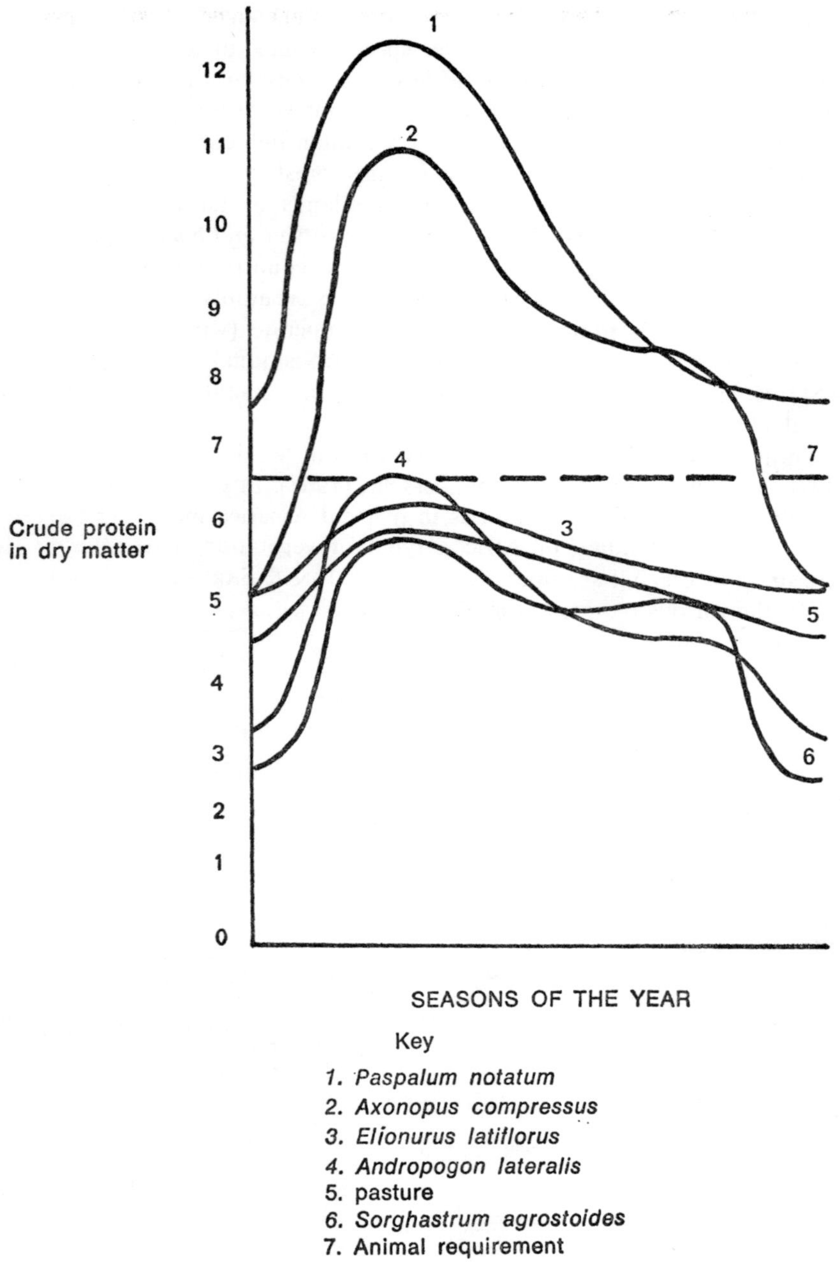

Fig. 3/6: Crude protein content of the forage of a natural pasture and of different species of grasses at different seasons of the year (FRETES, SAMUDIO & GAY, 1970).

4 SURVEY AND ASSESSMENT OF RESOURCES

4.1 The integrated or holistic approach

Specialists concerned with the evaluation of the total environment of tropical grazing lands accept the fact that their own detailed studies should be a part of or a sequel to an integrated survey, as defined by CHRISTIAN & STEWART (1968) – a survey of the physical and biological features, climate, water, soil, vegetation, minerals, terrain and natural environment. (See also annual reports of C.S.I.R.O. Division of Land Research.) Only indirectly are the most important natural resource factors considered, the human (rural) occupants of a country who are expected to follow up the recommendations arising from the survey, for the ultimate benefit of themselves and the urban populations who are dependent upon them.

'Of the physical resource factors, climate has the most direct significance for its general characteristics set the broad pattern of biological activity. Although land use may modify the microclimate, the broader characteristics of climate are not disturbed by man and the climatic pattern may be regarded as a permanent feature. The broad features of climate often change zonally in an orderly manner according to major global influences but this zonation is interrupted by topographic effects and the interrelations of land and water masses . . .'

'Each resource factor has its own spatial distribution and some have a temporal distribution as well. Thus, although the broader aspects of climate may vary geographically within a country, they will also differ from season to season . . . stream flows will vary seasonally according to rainfall, run-off and seepage. The grazing value of native vegetation such as natural grasslands will also be subject to season to season variations . . . Survey and assessment methods and techniques must be adapted to the nature and form of each resource factor'.

The Australian workers have adopted the term 'land' to denote the complex of climate, topography, soil and vegetation, since it is the combined effect of all and the interactions between them which are important. CHRISTIAN (1963) has stated: 'Land must be considered as the whole vertical profile at a site on the land surface from the aerial environment down to the underlying geological horizons, and including the plant and animal populations, and past and present activity associated with it. There are many features in this total profile, some easily observable like the soil and vegetation, some measurable such as the rainfall and surface slope, but many not so readily observed or measured . . . . The many features of this total profile vary from site to site and their many combinations and interactions result in a vast array of land types, each with its own potential and limitations for agriculture and forestry, each presenting its own specific barriers to the achievement of maximum plant or animal production. Rarely does one feature alone determine productivity. It is the combination of all that is important, and if we are to understand land we must think of it in terms of this complex rather than only of the individual components of it'.

The holistic and separationist approaches are discussed by ZONNEVELD (1972a) in Volume VII, Chapter VII; 4 Lectures on land science, land survey and land evaluation. For an international symposium, ZONNEVELD's discussion on the use and comparability of terms in different languages, and of the careless use of the word: 'ecosystem', is of particular value. Reference is made to the assertion that vegetation is a holistic expression of all environmental factors, that a vegetation classification unit is really a characterisation of the land as a whole, and can thus be used as the sole basis for land classification.

While agreeing with ZONNEVELD that such 'attempts are often too sweeping", one may make a case for specialized studies of vegetation in general, and of grazing lands in particular, to follow an integrated or holistic survey. After the more general survey has defined land within a given land system or unit as being actual or potential grazing land or sown pasture land, the subsequent detailed analysis and assessment of that resource are the job of specialists in botany, ecology and vegetation history and evolution.

Over vast areas of the tropics and subtropics, the vegetation as defined by the land scientists is rarely climax. The grass, shrub and tree cover on tropical grazing lands is already at a low stage in ecological regression, so much so that a semi-arid vegetation may be found growing in a subhumid climate, and grass savanna in a forest climate. Thus it is necessary to query the Australian statement quoted above, that land use may modify only the microclimate, in view of the presence of widespread desiccation of the total environment following haphazard or even planned devegetation for land 'development'.

The place of vegetation, grassland and land use in the overall definition of land systems is demonstrated in the survey made by the British Land Resources Division in North-East Nigeria (Vol. 3)(BAWDEN, CARROLL & TULEY, 1972), and in the accompanying statement (Vol. 4) on present and potential land use. A *land system* is here defined as an area or group of areas throughout which can be recognized a recurring pattern of topography, soils and vegetation. The components of this pattern are *land facets*. The land systems (123 in all) are grouped into *land regions*, and these again into *land provinces* (five defined). In spite of the failings of the land-system concept, it is regarded as admirably suitable for rapid reconnaissance surveys based on aerial photography.

In recognizing twenty grass cover communities, both physiognomy and species composition are closely related to the main agro-ecological zones. In the northern sector with a mean annual rainfall of 500 mm. or less, the short grass cover is typically annual (*Aristida, Chloris, Digitaria*). The few perennials are highly xerophytic in adaptation. In the central Sudan Zone, *Loudetia* spp. are conspicuous with shorter members of the Andropogoneae. When the rainfall exceeds 750 mm., the tall grass covers are dominated by annual and perennial members of the Andropogoneae.

4.2 **Significance of climate**

Throughout these Chapters, the influence of climate is repeatedly stressed as being more significant than geomorphology, geology or soils in governing the botanical composition of the ground cover in tropical grazing lands. Much of the area within the intertropical zone is characterized by an alternation of wet and dry seasons. It must be remembered that, in general, the grass species of undisturbed communities are fully adapted to the rigorous conditions and physiological stresses of long dry seasons. It is only where these conditions are exacerbated by extraneous factors introduced by man that species high in the grass cover succession suffer and are eliminated in favour of species more able to persist under conditions of 'management' – that is, following excessive year-round stocking and fires.

It is impossible here to review the climatic basis of distribution and behaviour of tropical communities and herbage genera and species; WALTER's studies (1964; 1970 and 1971b) are relevant. Perhaps one may be able to recognize distinctions between mesophytic and xerophytic environments, especially in relation to taxonomic advance and rate of speciation (9.4). Transition zones between one ecoclimatic region and another are of special interest in this respect. Species adapted to a monsoonal rather than a summer rainfall environment may be more aggressive and stronger in competition than species from summer rainfall or other, less exacting environments (5.4).

Some references to current studies will guide students to a wider literature, especially with reference to climatic analogues throughout the intertropical zone:

4.2.1 AUSTRALIA

The climatic factor in Australian grassland ecology – FITZPATRICK & NIX, 1970.

Climate in an integrated approach to ecology of arid Australia – Ross, 1969 – technique of detection of homoclimates (brigalow) – Russell & Moore, 1970 – comparative climatology of Australia and Argentina – Prescott, Collins & Shirpurkar, 1952.

4.2.2 Asia

Ecoclimates and bioclimates of monsoonal and equatorial Asia and Africa – Whyte, 1968a – occurrence of grasses and legumes in north-east Taiwan and Florida – Wang Chi-chu, 1959 – ecogeographic analysis of climate and vegetation in Sri Lanka – Mueller-Dombois, 1969 – bioclimates of Indian subcontinent and their analogous types in the world – Meher-Homji, 1963 – climatic analogues for the dry zone of India – Subrahmanyam & Sastry, 1969 – hydric balance – product of xero-thermic index and deficit of saturation – Legris, 1972.

4.2.3 Africa south of the Sahara

Agroclimatology of highlands of eastern Africa – Brown & Cochemé, 1969 – rainfall reliability: basic limitations to land use in East Africa – East African Royal Commission, 1955, chapter 20 – confidence limits for seasonal rainfall in Kenya – Woodhead, 1970; Woodhead, Waweru & Lawes, 1970 – rainfall-altitude relation and series of vegetation types in Kenya – Trapnell & Griffiths, 1960 – agroclimatology of a semi-arid area with monsoonal summer rains in Africa south of Sahara – Cochemé & Franquin, 1967.

4.2.4 America: South, Central and Caribbean

The tropical and borderline climatic provinces of Mexico – Contreras Arias, 1942 – climatology of South America – Eidt, 1968 – ecological subdivisions and climatological aspects – Schwabe, 1968.

4.2.5 Evapotranspiration

Studies of actual and potential evapotranspiration from various types of land and tropical crop and vegetation covers include Rijks & Walker (1968), Ojo (1969), Dagg, Woodhead & Rijks (1970), Dagg & Blackie (1970), and Cochemé & Franquin (1967). Oliver (1969) notes that there are two situations in the Sudan in which actual evaporation will depart from potential evaporation. In unirrigated, sparsely vegetated areas, actual losses will fall short of the estimates; from irrigated lands, especially small pump-irrigated schemes, calculated values based on the site climatic data can be considerably exceeded by actual evapotranspiration. Studies on potential evapotranspiration from grass covers are reported also by the Landbouwproefstation (Agricultural Experiment Station) of Surinam.

ELDIN (1970) has determined evapotranspiration in a grass cover (*Paspalum notatum*) at the O.R.S.T.O.M. station in Côte d'Ivoire, using a neutron gauge to record water profiles; the same worker (1969) has modified a method for measuring light energy under plant canopy, adapted to the study of radiations of small intensities and of special spectral compositions.

4.2.6 CLIMATIC BELTS

ANDREAE (1971) has related types of animal husbandry and the use of naturally available or purchased or cultivated feed and fodder resources to the types of climate (in belts – Fig. 4/1) and the availability of resources for development, in thornbush steppes and dry savannas. The

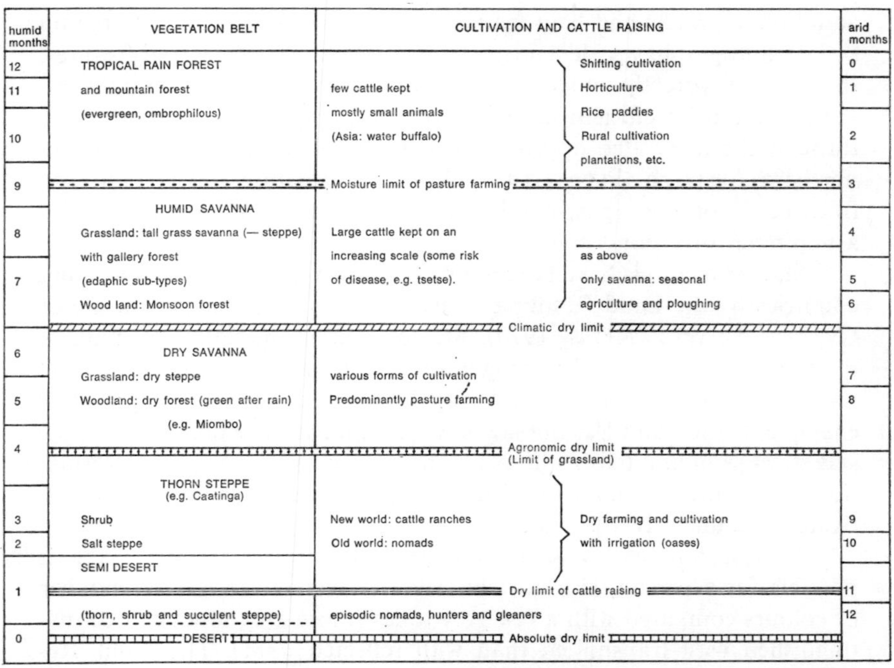

Fig. 4/1: The tropical climatic belts and their land-use systems (ANDREAE, 1966).

sequence is from nomadic and migratory grazing to the provision of additional watering facilities, and to introduction of the camp technique. Further intensification will depend upon the availability of capital for standard licks, supplementary feeding stuffs (concentrates, baled hay) and ultimately the cultivation of fodder crops on arable land.

4.3 Aerial photography

The use of aerial photographs, ground checks and subsequent interpretation is a fundamental technique in integrated surveys of land resources (Christian & Stewart, 1968; Bawden, 1965). The application of the technique to southern Asia is discussed by Verstappen (1969); in addition to the usual survey of geomorphology, geology, soils and water resources, reference is made to calculations of the relative period in areas under shifting cultivation and to the special features of central Rajasthan; widespread lime concretions at depths of 30 to 120 cm. prevent addition of rainfall to groundwater, necessitating the construction of 'tanks' for storage of surface-water; their siting has a profound influence on the sociology of the region.

The cost of survey can be reduced by interpretation from aerial photographs, by the establishment of a correlation between patterns on aerial photographs and land features (Perry, 1964). As the patterns on aerial photographs are dependent on the integration of all land features, they can be correctly understood only in terms of all these factors. Conversely, the most efficient use of aerial photographs can be made only through the integrated approach, e.g. for surveys of pastoral areas in northern Australia (Perry, 1967). Much information can be extracted from aerial photographs, making it possible to cover large areas with a minimum of ground control.

Similar methods have been adopted for surveys in African and other countries by the Land Resources Division of the British Directorate of Overseas Surveys (Rains, 1970). Stereoscopic examination of pairs of photographs facilitates the recognition of differences in land form, soil and vegetation, and boundaries may be marked on photographs. For example, in the Land Use Survey of Malawi, it has been possible to apply statistical sampling to air photo coverage, to provide adequate information economically, about the distribution of types of cultivated, uncultivated and uncultivable land.

New techniques involving colour and infra-red colour make it possible for a person with normal vision to distinguish a large number of colours compared with a few grey tones (more grey tones can be distinguished with transmitted than with reflected light). The Land Resources Division finds that results of analysing vegetation, soil and land-use patterns on true and infra-red colour air photographs are only marginally better than results from black and white prints. The use of infra-red colour facilitates remote sensing in the determination of the grass cover and/or the quantity of herbage present; it is also of value in range inventory studies. The recent development of high-altitude photography and the use of multispectral sensors to identify natural formations is unlikely to supersede conventional photography for some years (RAINS, 1970). Most workers will continue to use films recording the visible and near infra-red bands of the spectrum exposed at medium altitudes in fixed-wing aircraft.

Grassland, as defined by PRATT, GREENWAY & GWYNNE (1966) can be recognized on aerial photographs by its smooth grey textureless tone and its location (seasonally flooded plains and sites with impeded drainage, apart from high altitudes). Although grassland can be recognized and the density of tall shrubs and trees determined on 1 : 60,000 scale photography (RAINS, 1970), it is most difficult to assess the density of woody plants in low shrub savanna at this scale. The problem was resolved for resource surveys of Botswana (RAINS & YALALA, 1972) by analysing the distribution and type of the burning patterns. The characteristic burning pattern may appear as black or shades of grey. A pattern from fires over a period of seven years is an indication of good forage production. Patterns from deliberate burning suggest that deterioration is occurring. One must, however, be careful in interpreting fire patterns until the nature of the forage and density of the shrub can be checked.

The International Institute for Aerial Survey and Earth Sciences, Enschede, has reported on vegetation patterns in a savanna region of Northern Nigeria, during the UNDP/FAO survey of the soil and water resources of the Sokoto Valley. The interpretation of the relation between these and other patterns, clearly visible on aerial photographs, to the edaphic conditions played an important part in speeding up the reconnaissance survey (ZONNEVELD, DE LEEUW & SOMBROEK, 1971). Both the main patterns, the 'pseudo-dune and gully-pattern' and the 'black and white dot pattern' could be subdivided into sub-patterns in which various pattern 'elements' were distinguished.

ZONNEVELD (1972b) would agree with British workers that the ordinary, old-fashioned black and white panchromatic film will remain the cheapest and still the most valuable tool far into the future. Thermal sensing (far infra-red) will never be useful for general grassland surveys. Radar will certainly be interesting for small-scale survey, but so far there is little experience with it. The more conventional remote sensing like near infrared (preferably combined with visible radiation in false colour) is of particular value on vegetation types that have much open space, as in arid or semi-arid zones. The contrast between bare soil and patches covered with vegetation is sharp, and enables a good study of patterns to be made.

4.3.1 Skylab Data

FAO (1971) has prepared a working paper on the use of remote sensing and the use of spacecraft/skylab data for integrated survey of natural resources, and the evaluation of soil, water and forest resources. Reference is made to space photographic surveys in the Sudan, Colombia and the Philippines.

4.4 United Nations Programmes

4.4.1 International Programme for Grazing Lands

Following discussions between the International Biological Programme and FAO, a project has been drawn up for a programme of training, analysis and synthesis of data and distribution of information on the world's grazing lands (Unesco, 1971). It is considered to be unrealistic that countries should individually attempt to start separate, full-scale research programmes. U.N. Agencies are conducting studies and supporting training. The I.B.P. terrestrial productivity group has, through its grassland, tundra and arid land groups, started research on grazing lands, and established a network of communication (1.6). However, there has not yet been any integration, nor any detailed analysis and synthesis of results.

A systems analysis approach will be taken in the organization of data, information, experience, and ideas to develop models and analyses concerning important practical and theoretical problems. The programme will include development of a data bank of information on grazing lands and a programme of both informal and formal training for participants from developing nations.

Funds have been provided by the United Nations Development Program for a start to be made with the Mediterranean region, under the Man and the Biosphere Programme (1.6).

4.4.2 FAO Resource Survey in South America

A regional project for the evaluation of grassland resources has been in operation in the Regional Office of FAO in Santiago, Chile since 1970. The objectives for the project, which covers the tropical and temperate regions of Latin America from Mexico to Patagonia, are (Blydenstein, 1972): collection of basic information on the distribution and characteristics of grasslands,
description of correct systems of management,

identification of problems in optimal use of the resources,
establishment of priorities for research, and
rapid surveys of grazing resources in selected countries.

The methodology of the rapid surveys consists of the collection and manipulation of existing statistics on livestock production in relation to available information on grazing resources. Studies are completed or are in progress for Venezuela (BLYDENSTEIN, 1971), Peru, Chile, Argentina, Colombia and the Caribbean Islands. These can be no better than the statistics upon which they are based. Estimates are also obtained on grazing resources, extent of grasslands used, types of natural grassland or improved pasture, availability of additional forage and feed resources, carrying capacity with its seasonal fluctuations. A new design for an animal industry may be technically possible, but perhaps not economically feasible. For such integrated studies, the grassland work is coordinated with that of the FAO Livestock Development Group, also located in Santiago.

Numerous grassland surveys are concerned with the floristic composition and ecological status of the grass cover, whereas an evaluation of resources calls, according to BLYDENSTEIN, for a different set of data of which the botanical information is only a part. Yet it is admitted that it is the botanical and ecological factors which impose restrictions on the degree of livestock development. This is seen particularly in the seasonality of growth in the dry tropics. Seasonal variations in forage availability in the humid tropics are also important; growth of *Pennisetum purpureum* is depressed during the very wet season due to insufficient sunlight; *Digitaria decumbens* has two seasons of lower forage availability, one when it is in flower, another in the drier part of the year, even if rainfall then exceeds 50 mm. per month. Mineral imbalances are significant in livestock production throughout Latin America (12.5).

4.4.3 SURVEY IN KENYA

A UNDP/FAO Range Management Project became operational in Kenya in 1966, at the request of the Government. A credit of U.S. $7.2 million was also provided for rangeland development in specified areas, negotiated with the International Development Association and the Kingdom of Sweden. The three main objectives of the UNDP/FAO project are:

land-use survey and socio-ecological planning,
investigations on problems of forage and animal production, and
education of specialist staffs.

CASEBEER (1969) has described the plans for that part of the project dealing with integrated survey, through the six phases: review of available data and literature; reconnaissance; field studies; assessment and coordination; projections and recommendations; and reporting.

4.5 The approach of the plant sociologist

The Centre d'Etudes Phytosociologiques et Ecologiques at Montpellier has developed a method of identification inventory and mapping of vegetation and environment (LONG, 1966). Plant ecology is distinct from phytogeography, which is the study of natural vegetation or modified plant communities, when the units studied are large areas of physiognomic homogeneity, the 'plant formations' and 'vegetative types' of classical plant geography. Plant ecology has as its aim the simultaneous study of vegetation and environment, the main object being an understanding of the biological processes of ecosystems.

The first stage leading to the preparation of the large-scale maps required by range specialists is a sampling of plant communities in their natural habitat, using special forms on which are assembled the basic data, coded for transference to punched cards. DESCOINGS (1971) has discussed the use of standardized data sheets devised for analysis of the structure of intertropical herbaceous formations, including the formations classified in the Montpellier 'Code écologique' (GODRON *et al*, 1968) and especially those currently called savannas, steppes and pseudosteppes. Correlations between species and environmental factors are attempted and statistical tests of significance are made (LONG, 1966). The result is the establishment of 'ecological groups' which are indicators for the respective environments. These groups are used in the compilation of vegetation/environment maps which are accompanied by maps of the factors of the environment and of land occupation.

In a separate paper LONG (1969) discusses these techniques and their cartographical expression in greater depth. A general model of the perception of biogeographical facts and of their reciprocal relation is termed the pyramid of perception; this comprises five levels, from the small to the largest scales. The application of this concept to plant ecology and plant geography is important.

Efficient and objective sampling methods in areas used for grazing and browsing of wild and domestic livestock will, according to LONG, help developing countries to make up for lost time in the survey and analysis of their natural resources. The cost of phyto-ecological studies (1965 figures) including mapping but not printing, can vary from 4 cents U.S. per hectare for small-scale mapping of arid areas with open vegetation, to U.S. $2.00 per hectare for large-scale mapping in temperate regions with a closed vegetation.

The concepts and methods of plant sociology have been applied to the tropical conditions of Brazil (CAIN & CASTRO, 1959), in studies of the structure and distribution of vegetation as they relate to medical ecology (malaria, yellow fever, schistosomiasis), all limiting factors in the evolution of ecosystems based upon animal husbandry in forest/savanna environments.

4.6 Review of international literature

A review of the available world literature for the preceding twenty or thirty years, on methods used in quantitative ecological studies, and more particularly of grazing lands, was prepared by Brown (1954). Chapter 1 on vegetation and its sampling is subdivided into open grasslands, dense grasslands, shrub/open grassland, shrub/dense grassland, and dense shrub. Aspects covered include the theory of sampling, apparatus to be used, botanical analyses, measurement of yield and grazing capacity, estimate of cover (range survey), the animal as a measure of pasture production, the problems of utilization, reconnaissance surveys, and range condition.Because of the state of development at that time, there is little reference to tropical grazing lands.

4.7 Range condition analysis

This is considered as a tool for the study of unclassified grazing lands in general, and in Brazil in particular (Humphrey, 1966). The application of this North American technique requires a knowledge of the nature of ecological succession of vegetation and the potential of specific sites for forage production. Ranges are classified for their potential for maximum production into four classes: excellent, good, fair, poor, on the basis of five criteria: forage composition (botanical), ground cover, plant vigour, litter, erosion. Condition trend is also determined on the basis of criteria that indicate an upward or a downward trend, these being the kind of plants that are reproducing, their vigour and seed-setting characteristics, and degree of erosion.

In Rhodesia, it has been agreed (Southern Rhodesia, 1962) that the most important criteria for the classification of condition in African veld are the relative abundance of trees and shrubs, of perennial grasses, of annual grasses and herbs, and of bare ground. Because, to the grazier, the best veld is that which provides the most forage for domestic animals, especially cattle, year in and year out, the highest condition class is veld which consists of perennial grasses without trees or shrubs, with few annual grasses, and a minimum of bare ground.

African workers have to distinguish between a classification of condition class that could be used in a succession advancing towards bush, and condition as understood in the American range in a succession advancing towards grass. Condition in American range is usually related to the climax species present (76 per cent means condition excellent, for example). In Africa, the situation is reversed. In vegetation with 76 percent of climax species, condition for grazing is very low for anything other than the browsing species of game. Condition classification here, and probably in the Indian subcontinent as well, should be based on a 'theoretical condition of perfection, that is, on grass from which the trees have been excluded by management' (Southern Rhodesia, *op. cit.*; VINCENT, 1962; RATTRAY, 1962).

Little is known of the potential for forage production of many million hectares of brush- or tree-covered lands of South America (HUMPHREY, 1966); in Brazil, for example, is it possible to transform them to grass range, as has been done on the chaparral and pine/juniper woodlands of Arizona? The three types of brush land in the arid north-east of Brazil, for example, serido, sertão and caatinga, are very extensive and yet produce relatively little forage.

4.8 Grassland Survey of India

In 1953, the Indian Council of Agricultural Research decided to undertake a rapid reconnaissance survey of the grass covers of India (DABADGHAO & SHANKARNARAYAN, 1972; WHYTE, 1968a). Two members of the survey team (P. M. DABADGHAO & B. D. PATIL) received training under FAO fellowships in the United States of America and Great Britain, and others were trained at the Forest Research Institute, Dehra Dun. Up to that time, Indian and British botanists had done invaluable work on the recognition, naming and classification of the great number of grass species in the flora of the subcontinent, and N. L. BOR (1960) has produced a major work on the subject. Comparatively little attention had, however, been given to the grass communities, to their botanical composition, and to the nature of the succession that occurred under the influence of biotic factors.

The objectives of the survey and the techniques adopted on selected one-acre plots on over 500 sites throughout India, were:

a. Floristic composition
   (i) listing and classification of species
   (ii) estimates of percentage composition by the pace-transect method
b. Density of plant cover
   judged by the square foot density method from a circular sample enclosing an area of 100 sq. feet (STEWART & HUTCHINGS, 1936)
c. Forage production
   estimated by cutting and weighing the individual species from the 100 sq. ft. density circles

d. Plant vigour
assessed by the study of plant height, leaf length, length of seed-stalk, number of tillers, basal area and reproduction of species
e. Plant succession
data obtained by inference and by a comparison of floristic composition on undisturbed and adjacent overgrazed areas
f. Associated forest types
listing of tree species and classification of forest types
g. Soil types
data collected by methods adopted in soil surveys, for colour, texture, water-holding capacity, pH, total soluble salts, and soluble phosphate and potash.

4.9 Institut d'Elevage et de Médecine Vétérinaire des Pays Tropicaux, Maisons Alfort, France

The Institut d'Elevage et de Médecine Vétérinaire des Pays Tropicaux has evolved its own sequence of techniques for the study and mapping of tropical grazing lands (BOUDET & BAEYENS, 1963; BOUDET, 1966; BILLE, LEBRUN & RIVIÈRE, 1968/9; PAGOT, 1971):

a. detailed classification is made of phytosociological surveys of minimal areas located in different vegetation groups;
b. the floral components of each type of pasture so identified are obtained from a grid survey of a few examples of each characteristic group; a statistical survey of these results facilitates a more exact assessment of species frequency and basic cover if perennial species are abundant;
c. the forage value of each pasture type is determined by random sampling; aliquot samples are made to determine nutritive value and trace-element content; and
d. the survey is completed by the construction of a map based on aerial photographs.

In the Sahelian zone, this mapping permits the assessment of optimal head of cattle per region, the adoption of a hydrological programme in harmony with pasture potential, the establishment of a rotational system of grazing by regular closure of wells in relation to optimal season for grazing the dominant vegetation, and the probable development of pastures under intensive grazing, controlled again by closing wells at critical times.

In the Sudanian and sub-Guinean savanna, the maps show the stock-breeding potentialities of specific areas. The problems associated with the development of temporary pastures may be solved by assessment of productivity and control of pasture development by periodical grazing of 2,500 m^2. areas, and by making annual grid surveys.

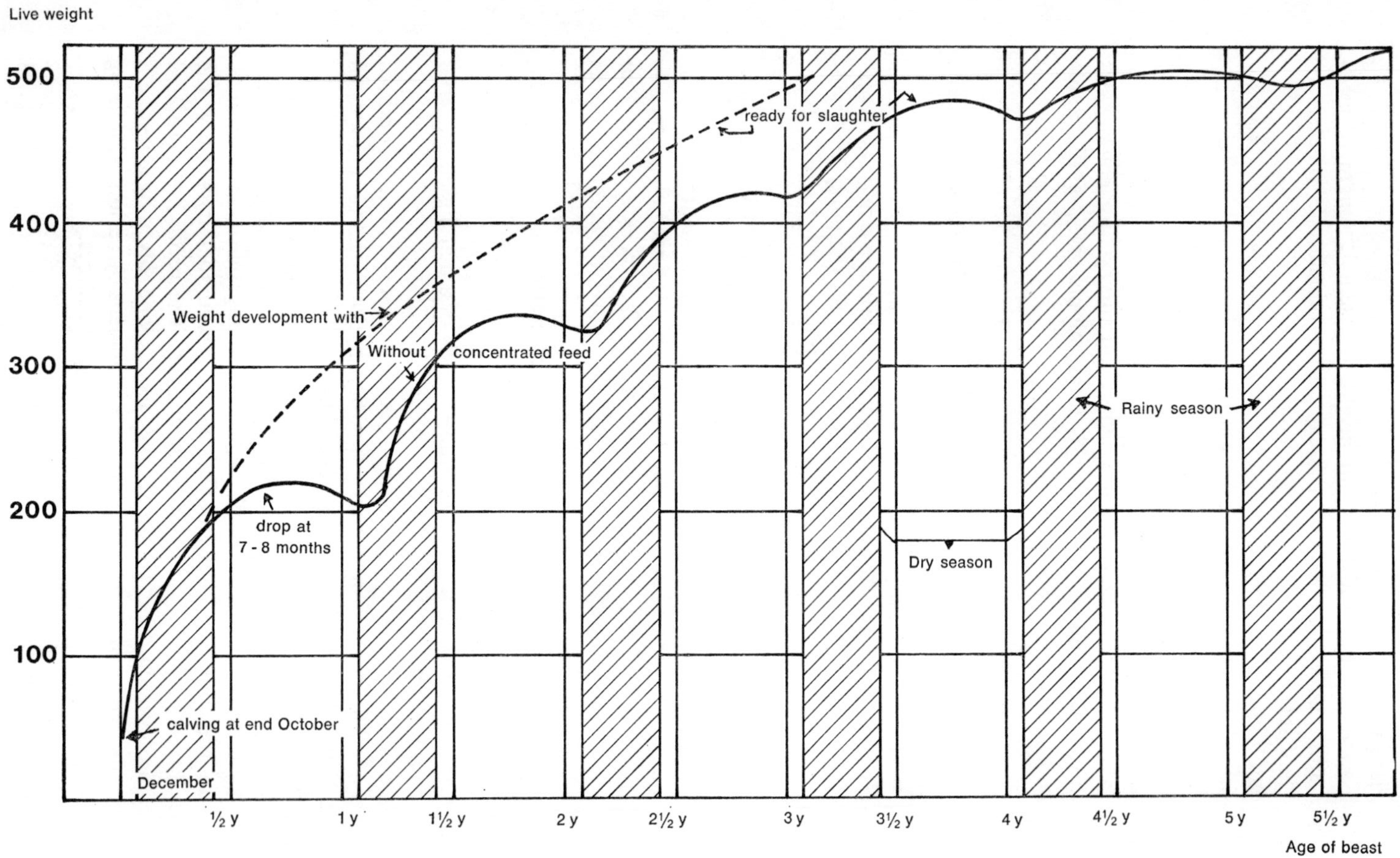

Fig. 4/2: Liveweight increase of young cattle in extensive grazing in central south-west Africa (ANDREAE, 1966).

BOUDET & ELLENBERGER (1971) describe the techniques used in an agrostological study in the Republic of Mali:

a. definition of natural pastures following inventory of abundance/dominance of species;
b. establishment of synoptic tables of the principal vegetation communities;
c. biological or growth forms of species with a significant distribution (mesophanerophytes, nanophanerophytes, geophytes, etc.).
d. relation between types of pastures, and the major physiognomic groupings recognized by TROCHAIN (1957):
   forêt dense sèche
   forêt claire
   savane
   savane boisée
   prairie (aquatic or marshy)
e. subdivision of types of vegetation other than forest according to the characteristics of the woody cover (arbustif, arboré, boisé), and
f. primary productivity of the herbaceous cover.

4.10 Assessment in terms of livestock production

No matter how well the human specialist may survey, analyze and assess the tropical grazing lands, the final arbiter is the animal. Thus the ultimate objective of the specialized techniques for surveying tropical grazing lands is the assessment of the resource in terms of actual and potential livestock production. At the same time, one must ensure the optimal conservation of soil and water, and the prevention of excessive desiccation of the environment, leading to reduced efficiency of rainfall. Soil specialists and vegetation ecologists state that the general trend in devegetation from tree formations to bush and scrub and to natural or artificial herbage communities is in the general direction of increased desiccation of the aerial and soil environments.

The final analyses in the resource survey will be of the quality, palatability and digestibility of animal food obtained from natural or artificial stands (Fig. 4/2). There is a considerable literature on the biochemistry of feed and fodder plants in different environments, on different soils, and with varying intensities of improved plant nutrition by the use of fertilizers or water for irrigation. A review of Australian work since 1962 on assessing potential intake when evaluating pasture species (the reason why an animal will eat as much as 50 per cent more of one cultivar or wild plant than it will of another with a similar chemical compositon) appears in Rural Research in C.S.I.R.O. no. 75, March, 1972 (Fig. 4/3).

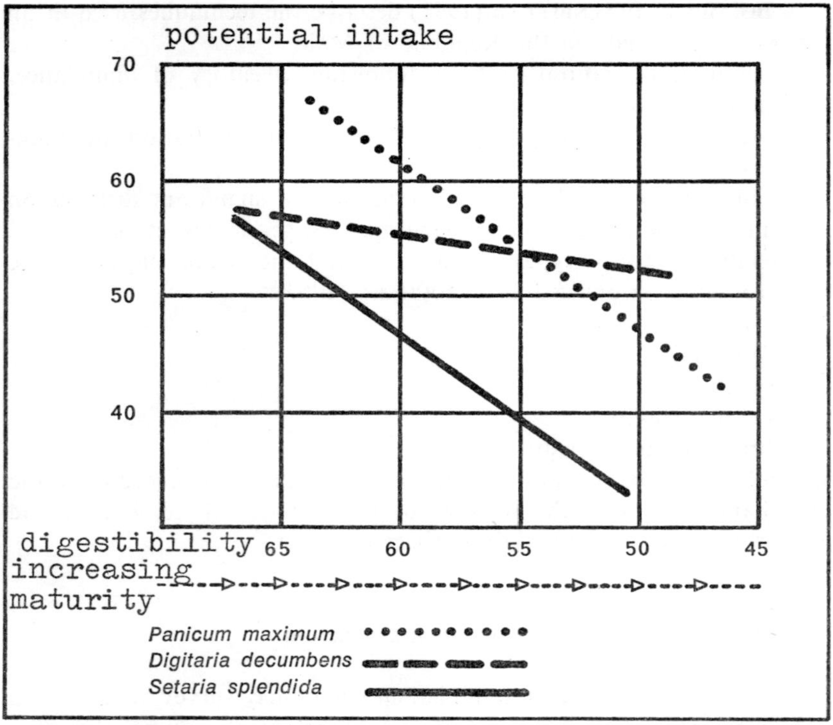

Fig. 4/3: Assessment of potential intake in evaluation of pasture species (*Rural Research in CSIRO* 75, March, 1972). Both digestibility and potential intake fall as a grass matures. However, the potential intake of *Digitaria decumbens* falls at a much slower rate than those of either *Setaria splendida* or *Panicum maximum.*

It is desirable to find methods for testing and screening plant species and cultivars for potential nutritive value and for digestibility at an early stage (Rural Research in C.S.I.R.O., June, 1971). This may be done by the use of equations (BUTTERWORTH, 1967; BUTTERWORTH & DIAZ, 1970), or by evolving an index for assessing palatability (BOURREIL, 1968). Special techniques may be adopted for the evaluation of tropical pastures in terms of animal liveweight gain, by using an animal latin-square design and variable and fixed-stocking rate designs (STOBBS, 1970). Fixed-stocking rate designs have more general application in the tropics, because they are easier to conduct; the results are repeatable and can be extrapolated to farm practice. Somewhat similar trials have been conducted by SCHALKWYK, LOMBARD & VORSTER (1968) on pasture on which *Themeda triandra* constituted about 50 per cent of the total basal cover, *Tragus koelerioides* 10 per cent, followed by *Eragrostis obtusa, Sporobolus discosporus, Cynodon hirsutus* and *Eragrostis chloromelas.*

Analyses of nutritive value and digestibility were combined with comparisons between quadrat and simulated grazing samples. In a series of studies at the University of the Orange Free State (last reference: COETZEE, LE ROUX, DU TOIT & VAN RENSBURG, 1970), investigations with tropical significance have been made on the relation between fodder and ruminal composition and its possible application to the nutritive evaluation of a relatively pure stand of *Themeda triandra*.

The practical techniques adopted by the Institut d'Elevage et de Médecine vétérinaire des Pays tropicaux for the assessment of tropical pastures through forage analyses have been described by BOUDET & RIVIÈRE (1968). On the analogy of the livestock unit or l'unité gros bétail (U.G.B.) of temperate countries, a reference animal of 250 kg. can be taken as the livestock unit in tropical countries, l'unité bétail tropical (U.B.T.). The theoretical daily consumption has been estimated at 2.5 kg. of dry matter per 100 kg. of body weight; the needs of U.B.T. are referred to the kilogramme of ingested dry matter as equivalent ration expressed in fodder unit and digestible nitrogen. Analyses of the feeding values of over 500 tropical fodder plants showed: the importance of woody pastures as supplementary feed, in digestible nitrogen during the dry season in the steppe and the savanna, the periodical importance of some supplementary species, including *Crotalaria podocarpa* and *Alysicarpus ovalifolius*, for feeding cattle in the steppe, the limited number of species which can ensure maintenance requirements over a long period in the steppe, and the importance of the time of growth of perennial species in the savanna, since the feeding value of the regrowth depends more upon this factor than on time of harvesting.

5 SYNECOLOGY AND SUCCESSION

5.1 Percentage of land under extensive and intensive use

In making a resource assessment of tropical grazing lands, it is essential to accept the fact that, for technical and economic reasons, only a very small percentage of the vast area of land involved can be converted to highly productive pastures. A further percentage, still not by any means large, might be improved at lower cost under the appropriate socio-economic conditions, by reseeding following ploughing, or by the surface introduction of grasses and especially legumes into the existing sward. Fencing would then appear to be essential to provide the control of the harvesting mechanism in the ecosystem – the grazing animal or the cutting knife – and thereby to ensure the persistence of the synthetic plant communities so created.

Over the greater proportion of the tropical grazing lands, however, only those ecological and related types of management discussed in Chapter 6 are applicable. Those who would use these methods should therefore be fully informed of the principles and practice of succession and synecology in the widest sense, in so far as this branch of vegetation science relates to those communities which are required to provide sustenance for the wild and domestic livestock in diverse biological and economic ecosystems. It is realized that it is ecologically heterodox to speak of succession in one component only of a complex biological ecosystem of tree/shrub/gramineous ground cover, namely the grass communities, but it has its practical advantages.

Some idea of the areas involved are quoted by Crowder (1971) for Latin America, on the basis of FAO figures:

	Grasslands classified as natural	
	Million hectares	Percentage of total
Brazil	125	93
Peru	27	93
Ecuador	4	90
Bolivia	11.3	90
Venezuela	16.7	90

5.2 Grass covers in forest climaxes

There are no climax grass covers in the intertropical zone, except perhaps in small and localized areas of climatic or edaphic climax. Thus, the grass cover ecologist is concerned with what may be called ground cover subclimaxes in a wide range of shrub and tree formations; most of these are themselves subclimaxes in regressive forest succession, due to the operation of anthropogenic factors. This was recognized in RATTRAY'S map of the Grass Cover of Africa (1960), in which the existing tree cover was introduced as a backdrop to the grass map; also by the team conducting the Grassland Survey of India, who always noted the tree species and forest types when recording the total site cover (4.8).

Some writers have suggested that there is a degree of parallelism between forest type and the associated ground grass cover (SHANKARNARAYAN, 1963, Fig. 5/1, WHYTE, 1968a). Grassland workers are fully aware of the type of tree and shrub species that they find regularly associated with their grass covers, without necessarily referring to this as parallelism. It may be true to say that, in sites undisturbed by anthropogenic influences, tree canopies and grass ground covers are parallel, in that one would expect to find certain grass species and, in the more open

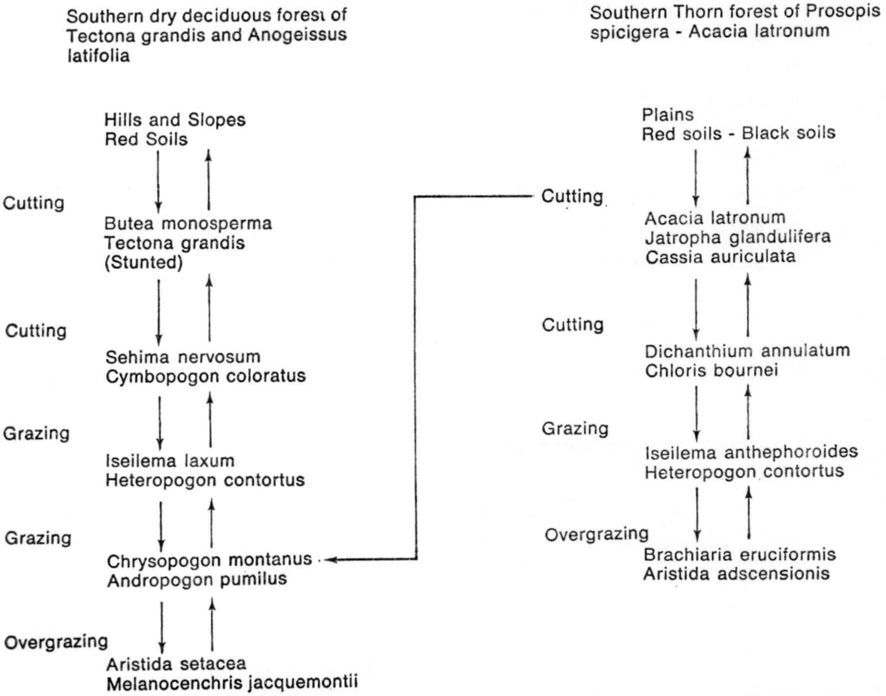

Fig. 5/1: Succession in two types of vegetation at Bellary, Mysore, India (SHANKARNARAYAN, 1963) – see also Bhatnagar, Fig. 5/6.

forests, grass communities associated with certain tree or forest types. When either the forest covers or the grass covers or both come to be used and misused, that degree of parallelism may be lost due to differential rates of regression.

Some examples of tree or shrub/grass cover associations for two of the vegetation zones south of the Sahara are given by BOUDET & BAEYENS (1963). It is to those types of vegetation that the sylvo-pastoral methods of husbandry are applied (6.8).

Steppe zone – Mauritania:

a. *Zizyphus mauritiana* and tufts of *Cymbopogon schoenanthus.*
b. *Acacia seyal* dominant, *A. raddiana, Balanites aegyptiaca* and ground layer dominated by *Andropogon gayanus* var. *genuinus.*
c. *Acacia seyal* and *Cordia gharaf,* tufts of *Cymbopogon schoenanthus,* some *Schoenefeldia gracilis*
d. trees of *Acacia nilotica* var. *tomentosa,* shrubs of *Mitragyna inermis* and *Feretia canthioides,* with *Oryza breviligulata* and *Echinochloa colonum.*

Sudanian zone – Mali:

a. shrubs of *Parinari curatellaefolium* and *Pterocarpus lucens* with *Ctenium newtonii.*
b. forêt claire with *Pachystela pobeguiniana, Bombax costatum,* with dominant shrubs *Acacia macrostachya, Boscia angustifolia* and *Combretum molle.*
c. very open tree layer with *Parkia biglobosa* and *Butyrospermum parkii* (species protected for their fruits), a shrub layer of *Terminalia avicennioides* and *Guiera senegalensis,* wooded savanna cultivated with sorghum and groundnut, with *Andropogon gayanus* and *A. pseudapricus* on old fallows.
d. rare *Terminalia macroptera* with *Hyparrhenia rufa* and *Brachiaria fulva.*

5.3 Authenticity of indigenous species

It has been proposed in Chapter 2 that many of the grass communities of the intertropical and especially the equatorial zone are composed entirely or partly by species which are not indigenous to that site. From the academic, if not necessarily from the practical point of view, it is desirable to classify grass covers into those that are truly indigenous and those which may be called derivative or evolved. In those parts of the intertropical zone formerly and until quite recently, in terms of geobotanical time, covered with tropical rain forests and other types of forêts denses, the few truly indigenous grasses to be found in their ecological niches could not have come out into the light and formed grass communities. Those heliophilous types that now exist in the cleared or burnt-over land must have come from elsewhere. A number of species may have arrived from different habitats and are meeting for the first time in new conditions of environment and interspecific competition.

Thus in south-east Asia and southern China species from Africa and the Indian subcontinent are found taking over after forest clearance where the few local indigenous and endemic species were unable to do so. In Latin America we have the phenomenon which PARSONS (1970) has called the 'Africanization' of the tropical grasslands of the New World, either by intentional introductions, or by accidents associated with two major events in American history, the slave trade (HOLDRIDGE, 1947), or the introduction of the Zebu cattle and the Asian buffalo. In Central America, over 75 per cent of the total pasture area is planted to, or has been colonized by introduced African species (HORRELL, 1972; BAZAN, 1972). *Hyparrhenia rufa* accounts for 70 to 80 per cent, or 3.5 to 4 million hectares of this total, on infertile land and under the local type of ranching; *H. rufa* is the only herbaceous plant in the derived savanna of Costa Rica that prospers under annual burning (DAUBENMIRE, 1972a); *Panicum maximum*, 5 to 15 per cent on more fertile soils; *Brachiaria mutica* 5 to 15 per cent on waterlogged sites; also *Pennisetum purpureum*, *Melinis minutiflora*, *Eriochloa polystachya* and, latterly, *Digitaria decumbens*.

Within Africa itself, one can trace a trend of species into the areas of climax forest as recognized by AUBREVILLE (1962 and chapter 3), from two directions in respect to their mesophytic and xerophytic origins. The mesophytic perennial species would progress from their ecological niches in the tropical forest zone (coastal and river flats, etc.) or from the Central African lakesides and marshes (THOMAS, 1966) into cleared forest land, to the limits of their genetical tolerance of aridity. Conversely, from the zones of the dry forêt claires came the xerophytic perennial grasses, advancing into the cleared forest zone to the limits of their genetical tolerance of humidity. Subsequent speciation and intercrossing in those new habitats has provided the gramineous flora of tropical Africa which is found today.

5.4 Relative aggression

Grass species of various origins within the intertropical belt exhibit differential aggressiveness. When African grasses were introduced intentionally or adventitiously into Central and South America, and where they found the environment to their liking, they entirely eliminated whatever indigenous grasses may have existed on the cleared forest land. The Indian and African grasses had virtually no competition from other grasses when they invaded south-east Asia, and until they began to compete with each other. When mesophytic African grasses were introduced into the Indian subcontinent, they were generally and still are 'cultivated' under rather special conditions, protected from competition with the Indian grass flora. But when dryland African grasses such as *Chloris gayana* and *Cynodon plectostachyus* are introduced into the Indian subcontinent, they are unable to compete with and are duly eliminated by the Indian species. Thus degree of competitive ability is governed at least partly by environment of place of origin; a grass that has evolved and can persist in its own rugged monsoonal, semi-arid environment is superior in that respect and in that environment to grasses from easier, summer-rainfall habitats.

5.5 Relative place in succession

If certain grass species are found to be low in ecological succession in one part of the tropics, does this mean that they will be found in a similar position in successions elsewhere? Species of the genera *Aristida* and *Eragrostis* are low in succession in the grass covers of the Indian subcontinent; it is known that, following protection from regressive ecological influences, these covers will proceed upwards through several stages to one of the few recognized sub-climax grass cover types with little or no *Aristida* or *Eragrostis*. Reports of vegetation surveys from Africa contain references to or show illustrations of grass stands dominated by different species of these and other genera. Can one extrapolate by suggesting that these African stands have been long exposed to similar regressive influences, and that they also would progress upwards towards a higher grass cover type, provided that these influences were removed, and also that seed sources of the ecologically superior species were still available? Critics of the method of ecological manipulation refer to the slowness with which this change takes place, but this is not by any means always so.

5.6 The pantropical ubiquitaries

Imperata cylindrica, Cynodon dactylon, Heteropogon contortus – these are the star performers for migration and aggression throughout all or much of the tropics. They call for a special study, involving geobotanical and ecological history, the latter in relation to anthropological factors, trade contacts and the zoological promoters of plant migration.

HARLAN, DE WET & RAWAL (1970a) have stated, with reference to the eight species of *Cynodon* L. C. Rich that they: 'fall clearly into four groups according to geographical distribution: South Asia and Indian Ocean – South Pacific Islands (*C. arcuatus*, *C. barberi*); East Africa (*C. plectostachyus*, *C. aethiopicus*, *C. nlemfuensis*); South Africa (*C. incompletus*, *C. transvaalensis*); Cosmopolitan with endemic varieties (*C. dactylon*). The fragmented geographic patterns, with limited distribution and complete genetic isolation of some species, and the narrow endemism of some varieties of *C. dactylon*, all imply a considerable antiquity and long evolutionary history' (see also Table 1 in HARLAN 1970).

The same authors (1970b) describe plants of a seleucidus race of *C. dactylon* var. *dactylon* as being strikingly coarse, vigorous, bluish in colour, often hairy, strongly winter-hardly, spreading aggressively by coarse stolons and rhizomes that are interconvertible.

'The center of distribution corresponds closely to the original Seleucid Empire (Pakistan to Turkey), and germ plasm from the race has infiltrated into eastern and southern Europe. It is probably an introgression product involving *C. dactylon* var. *afghanicus* and a temperate race of var. *dactylon*'.

It is not known whether there has been a consensus of opinion on the taxonomic age, possible centre or centres of origin and history of distribution of *Imperata cylindrica* (and the other named species of this genus of doubtful taxonomic status). SANTIAGO (1966) refers to complex interactions between aerial and subterranean parts and between genotypic and environmental variations, promoting the evolution of existing ecotypes and of new ecotypes. Swards dominated by *Imperata* frequently cover vast areas of land, for example, in the green cogon deserts of insular south-east Asia. They may become ecologically permanent at this low level of succession, when the controlling factors remain in operation, or in the absence of seed sources of species higher in the succession. With considerable labour and expense, agronomists plough and sow or plant superior species, which, however, have to be very vigorous and well-managed to prevent regrowth of the *Imperata* – for example at Keningau, Sabah, and at several centres in the Philippines. KELLMAN (1969) has compared *Imperata* grass covers with other types of vegetation and crop covers in respect of run-off, sediment loss and mineral loss from soils under shifting cultivation.

There are several reports of succession away from the *Imperata* phase. In the Sudano-Zambezian and Guinean areas of South Kasai and Katanga, RISOPOULOS (1966) reports that *Imperata cylindrica* rapidly becomes dominant after cropping; the species colonizes adjacent areas by its particularly dense rhizomatous root system, and is well adapted to withstand bush fires. In some sandy areas, one finds the tall caespitose, less aggressive species, *Urelytrum thyrsioides*. If a bush fallow is not brought back to cultivation, species of *Hyparrhenia* appear: *H. rufa*, *H. filipendula*, *H. dissoluta*, followed soon afterwards by *H. familiaris*, *H. diplandra*, *Panicum phragmitoides* and *Elymandra androphila*. In the least humiferous soils, *Loudetia arundinacea* and *Andropogon shirensis* predominate.

It has been said that, due to shifting cultivation, 40 per cent of the Philippine archipelago is covered by grass (predominantly *Imperata cylindrica* and *Saccharum spontaneum*, but never together), 16 per cent secondary growth forest, and only 10 per cent cultivated (WHITFORD, 1906). There is little doubt that most of the grassland areas in the Philippines, ranging from treeless through various stages of regeneration, are predominantly *Imperata* grasslands (RATTRAY, 1969) which have developed following shifting cultivation, assisted by fire. In the early stages of succession, common grasses of undisturbed areas such as *Paspalum conjugatum*, *Digitaria sanguinalis*, *Eleusine indica*, *Cynodon dactylon* and *Chrysopogon aciculatus* are found, but it is not long before *Imperata cylindrica*, *I. exaltata* or *Saccharum spontaneum* appear and eventually take over completely. Small areas of *Themeda triandra* have been noted, of doubtful ecological status and place in succession. *Imperata* seems to favour moister conditions and is less widespread in the drier parts of Luzon. Extensive *Imperata* grasslands are found in the islands of Masbate and Burias, in the west of Mindoro, and the central part of Bohol; in Panay; much of Negros Oriental; and most extensive of all in Mindanao, provinces of Bukidnon and Cotabato.

Fire-resistant plants with large underground structures (the orchid *Eulophia exaltata* and the shrub *Blumea balsamifera*) and the fire-resistant trees (*Antidesma ghesaembilla*, *Bauhinia malabarica* and *Acacia farnesiana*), can gradually eliminate the grass in their immediate neighbourhood (BROWN, 1919). When fire is excluded from a grass area, shrubs, vines and small trees can invade, leading in a few years to second-growth forest of small, rapidly growing trees. But the shade-intolerant *Imperata* and *Saccharum* are first replaced by other grass species forming taller, less dense stands.

5.7 **Indicator value**

The gramineous species of tropical grazing lands have considerable value as indicator plants. An experienced observer can find evidence in the occurrence of a certain genus or a certain type of grass cover for differing combinations and intensities of management practices (grazing, cutting, burning) and of burning with and without shifting cultivation; also for different soil types, soil moisture status, microclimate, presence of sub-surface hardpan, and so on.

Gramineae also play an important part as members of the distinctive floras of heavy-metal-bearing soils. DUVIGNEAUD (1958) and DUVIGNEAUD & DENAEYER DE SMET (1963) recognized types within the Katangan copper flora on an ecological basis – cuprophytes and plants growing on soils with the highest copper values, and which may sometimes be restricted to copper soils. Gramineae and Cyperaceae predominate in various life forms on soils with the highest values (5,000 to 10,000 ppm.) – *Eragrostis boehmii*, *Monocymbium ceresiiforme* and *Andropogon dummeri*. Cuprifuge species, those which never occur on copper and by their absence indicate its presence, include species of *Hyparrhenia*. WILD (1969a) has compared the copper flora of Rhodesia with that of Katanga; the ecological pattern rather than the individual species may be used for the identification of copper soils, and include the short graminoid zone sedges and the grasses *Loudetia simplex*, species of *Andropogon*, *Danthoniopsis*, *Microchloa*, *Schizachyrium* and *Schmidtia pappophoroides*. WILD has also studied graphitic soils (1969b) and nickel-bearing soils (1970). The nickel flora is similar to the serpentine flora (WILD, 1965); the genera *Acacia*, *Aristida*, *Combretum* are common on both; the familiar Leguminosae, Gramineae and Compositae are equally important on both. The most frequent grasses are *Loudetia simplex*, *Andropogon gayanus*, *Aristida leucophaea* and *Danthonia intermedia*.

In the cerrados of Brazil (3.2.3), there is a significant negative correlation between aluminium and the critical nutrients Ca+Mg and K. This suggests that aluminium toxicity may occur in these senile latosols which average 75 ppm. Al, and that cerrado plants must be extraordinarily Al-tolerant (GOODLAND & POLLARD, 1973). The world flora contains a small, taxonomically unrelated group of plants that actually accumulate aluminium to several thousand ppm. in their leaves, dry weight. Aluminium accumulators contribute greatly to cerrado biomass, and are actually dominant in over 30 per cent of the stands.

5.8 Types of grass covers

Most of the abundant literature on the ground cover components of the vegetation on the tropical grazing lands represents studies in static ecology. Types of grass covers are classified in relation to ecoclimatic zones, edaphic conditions or altitude. Only rarely is any indication given of the nature and trend of succession within an individual type.

It is, therefore, not possible to say whether the grass cover type as recognized and described is the true grass subclimax for that particular site in that particular type of present or former forest climax.

When one has begun to look at grass cover types in the field and attempted to diagnose their place in succession, it becomes obvious that many if not most of the types now in existence are seral stages in a succession, or rather regression, caused by the operation of one or more of the major anthropogenic factors: fire, grazing, clearing for cultivation. Thus one comes to look with ecological suspicion at some of the grass cover types recognized by those who map merely what is there now. As already stated, the dominance of certain genera suggests a low place in the succession, as does a high percentage of annuals, because there are no undisturbed subclimax associations in which annuals provide a significant part of the cover: other species are indicators of excessive or badly-timed overgrazing, or of excessive or badly-timed burning, or of a combination of extreme pressures of shifting cultivation, burning and concomitant loss of soil fertility over a long period.

5.8.1 Australia

The tropical and subtropical grasslands of Australia are described in the special volume, *Australian Grasslands* (Moore, 1970), prepared for the XIth International Grassland Congress. For the grazing lands of this region, which comprise the grassy understorey of the woodlands which have been modified to a greater or lesser extent by the influence of man and his grazing animals, Shaw & Norman (1970) quote Perry (1960) and other publications of the C.S.I.R.O. Division of Land Research with reference to the Northern Territory and Western Australia. Perry (1960) records under each type of pasture the actual or possible reaction to grazing, leading to the increase or decrease of desirable or undesirable species.

In the Tropical Tallgrass of Queensland (see Chapter 6 of Moore, 1970), *Heteropogon contortus* is now dominant, with *Themeda australis, Bothriochloa bladhii* syn. *B. intermedia*, *B. ewartiana*, *B. decipiens* and *Aristida* spp. *Themeda triandra* may have been the original dominant. Botanical composition varies considerably, even over short distances; other genera include *Chrysopogon*, *Cymbopogon*, *Chloris*, *Dichanthium*, *Eriachne*, *Schizachyrium*, *Sorghum* and *Eragrostis*. The species composition of this grass cover on any specific site is due to edaphic and use factors. *Themeda australis* and *Heteropogon contortus* are also found in the *Melaleuca* woodlands to the south and east of the Gulf of Carpentaria and in Cape York peninsula.

Cull & Ebersohn (1969) have studied the dynamics of semi-arid plant communities in western Queensland, more especially in relation to population shifts of two invaders, *Cenchrus ciliaris* and *Heteropogon contortus*.

The location of the main Laboratories and field stations of the CSIRO Division of Tropical Pastures in Queensland is shown in Figs. 5/2 and 5/3.

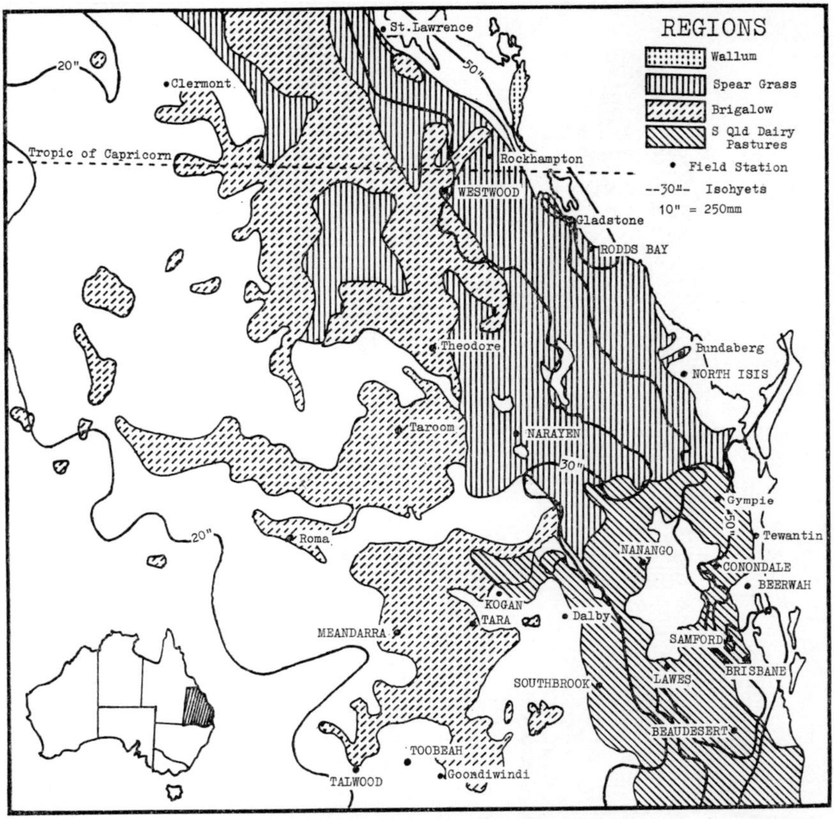

Fig. 5/2 Pasture environments and field stations where CSIRO Division of Tropical Pastures is conducting research based on the Cunningham Laboratory, St. Lucia.

5.8.2 Papua New Guinea

Studies on grass cover ecology relate to the monsoonal savannas at lower elevations and to the significance of the high-altitude grasslands. Earlier work by C.S.I.R.O. Division of Land Research is reviewed by Whyte (1968a).

Of the 255 genera in the flora of semi-deciduous forest and scrub and of eucalypt savanna in the Port Moresby area, 8 per cent are centred in Australia and/or the Pacific, 12 per cent in Asia and/or Malesia, and 80 per cent occur in and often beyond Australia and Malesia (Heyligers, 1972). Monsoonal conditions have not led to a stronger representation of Australian elements. The large percentage of widespread genera will decrease if one considers the whole of monsoonal south-eastern Papua; a comparison with Pacific and Malesian islands shows that the relation between the numbers of widespread genera and the total

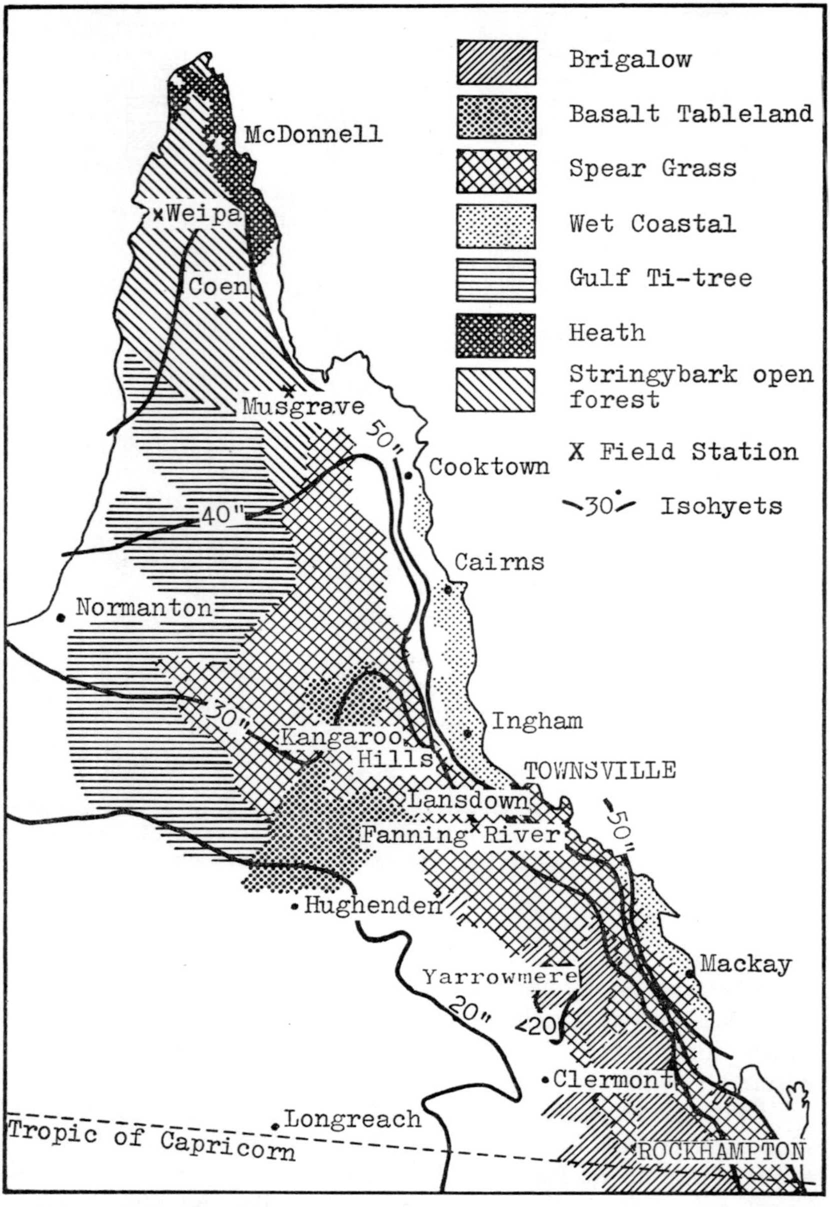

Fig. 5/3 Pasture environments and field stations where CSIRO Division of Tropical Pastures is conducting research based on the Laboratory at Townsville.

number of genera can be expressed by a logarithmic series. It is surmised that most of the widespread species reached south-eastern Papua before the Pliocene via drier areas on the Asian and Australian continents.

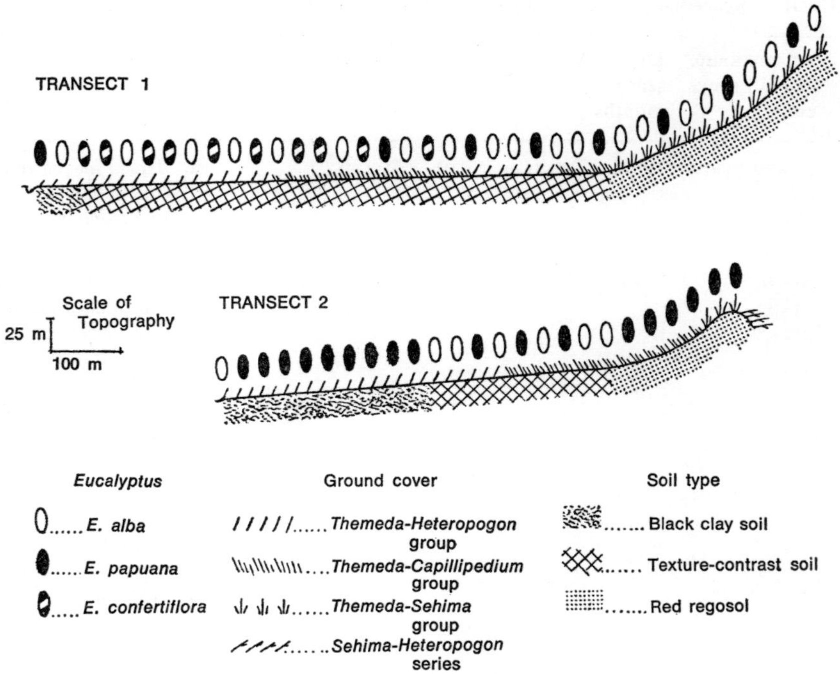

Fig. 5/4: Transect in a *Themeda australis/Eucalyptus* savanna, in Papua, showing the distribution and mutual proportions of the *Eucalyptus* species, and the distribution of the ground cover groups and of the soil types (HEYLIGERS, 1966).

In Papua, a *Themeda australis/Eucalyptus* savanna has a tall stratum of one or more species of *Eucalyptus*, some scattered *Albizzia procera*, and a grass cover dominated by *Themeda australis*, with *Heteropogon contortus* commonly associated, especially on stonier soils (see Fig. 5/4). Fires are common throughout the dry season, at the end of which most of the ground cover has been burnt.

JOHNS & STEVENS (1971) have listed the (mostly temperate) grasses of the national park on Mt. Wilhelm (4,510 m.). PAIJMANS & LÖFFLER (1972) have studied the high-altitude forests and grasslands of Mt. Albert Edward, in the Owen Stanley Range, the mountain backbone of East Papua. The major dominant on the most common soils, relatively deep and well-drained, is *Danthonia archboldii*, with local co-dominants *Danthonia vestita* and *Deyeuxia* spp. Some 70 per cent of the summit plateau would still be under forest, but for destructive fires associated with hunting.

The gramineous flora of these mountains should be considered alongside those of the arc of high mountain areas lying along the western Pacific (WHYTE, 1972b, 1973; LIU, 1971a).

Table 5/1 INDIA: Succession on a rainfall gradient, of vegetation, forests, grass covers, and types of

Zones	Approximate rainfall range, cm.	Dry season, months	Vegetation	Forest formation	Chief tree species
Arid	30 and below	10–11	Desert	Scrub	a. *Acacia jacquemontii* *Calligonum polygonoides* *Calotropis procera*
Semi-arid	30–60	7–9	Thorn forests	Open scrub woodland	(i) *Prosopis spicigera* *Salvadora oleoides* *Acacia leucophloea* *Acacia arabica* *Capparis decidua* *Balanites aegyptiaca* *Zizyphus nummularia*
					(ii) *Acacia senegal* *Anogeissus pendula* *Cordia rothii*
					(iii) *Acacia leucophloea* *Acacia arabica* *Acacia latronum* *Capparis* and *Zizyphus* spp.
					(iv) *Acacia planifrons* *Acacia latronum* *Albizzia amara*
				Savannah woodland	(v) *Pterocarpus marsupium* *Anogeissus latifolia* *Acacia latronum* *Albizzia amara*
Dry sub-humid	60–120	6–8	Woodlands, savannahs	Open dry deciduous types (i) Dry teak	*Tectona grandis* *Adina cordifolia* *Emblicata officinalis*
				(ii) Dry *sal*	*Shorea robusta* *Pterocarpus marsupium* *Terminalia tomentosa* *Anogeissus latifolia*

crop and animal husbandry (compiled by Y. SATYANARAYAN)

Grass cover	Main types of land-use	Main crops	Livestock	State
Lasiurus sindicus–Cenchrus spp.–*Panicum* spp. *P. turgidum* and *P. antidotale*)	Nomadic and semi-nomadic grazing, dry farming	Millets, sorghum barley	Camels, sheep goats, cattle and donkeys	Rajasthan
Cenchrus–Dichanthium annulatum, Eremopogon foveolatus, Eleusine compressa–Dactyloctenium sindicum, Aristida	Nomadic and semi-nomadic grazing, intensive dry farming, utilization of trees and shrubs for grazing, also cultivation with well irrigation, dairy industry (northern India), fallows of 3 to 7 years' duration	Lesser millets, *(Eleusine)* millets, sorghum, barley, cotton, oil seeds, some wheat	Camels, sheep, goats, cattle some buffaloes donkeys (note: goats are for mohair and mutton)	Rajasthan, Punjab, Gujarat, Maharashtra and Madhya Pradesh
Cymbopogon jwarancusa–Sehima nervosum	Intensive dry farming, multiple cropping, free grazing, utilization of forests and minor products for fuel and timber, fallows of 2 to 3 years (southern India)	As above, besides groundnuts, cotton, irrigated rice, sugar cane, chillies, wheat and pulses	Sheep, cattle goats, some buffaloes	Maharashtra, Mysore, Andhra Pradesh, Tamil Nadu
Dichanthium annulatum–Themeda quadrivalvis, Pseudanthistiria heteroclita				
Heteropogon contortus, Aristida redacta, Aristida funiculata				
Cymbopogon coloratus–Themeda triandra				
Cymbopogon coloratus Heteropogon contortus–Themeda triandra, Iseilema laxum, Chrysopogon montanus, Aristida spp.	Dry and irrigation farming, seasonal fallows, horticultural crops, utilization of forest products, forest grazing for local consumption and export	Millets, sorghum maize, cotton tobacco, oil seeds, wheat, pulses, mangoes, citrus	Cattle, buffaloes, goats, some sheep, horses, donkeys	Punjab, Madhya Pradesh, Gujarat, Maharashtra, Andhra Pradesh, Tamil Nadu, Uttar Pradesh
Chrysopogon montanus Bothriochloa intermedia Eulaliopsis binata				

Table 5/1 (continued)

Zones	Approximate rainfall range, cm.	Dry season, months	Vegetation	Forest formation	Chief tree species
Moist sub-humid	120–160	5–8	Open woodlands with eastern areas covered by *sal* and western areas by teak and *Terminalia*	Seasonal forests with mixed evergreen and deciduous species	*Tectona grandis* *Terminalia crenulata* *Terminalia paniculata*
				(i) Moist teak	*Terminalia bellerica* *Dalbergia latifolia* *Lagerstroemia lanceolata* *Grewia tiliaefolia*
				(ii) Moist *sal*	*Shorea robusta* *Lagerstroemia parviflora* *Dillenia pentagyna* *Pterocarpus marsupium* *Syzygium cumini* *Schima wallichii* *Terminalia tomentosa*
Per-humid	160–200	2–6	Moist forest at low and medium altitudes	Evergreen *seasonal* forest (western form)	*Actinodaphne angustifolia* *Olea dioica* *Pouteria tomentosa*
				(eastern form)	*Cinnamomum cecidodaphne* *Amoora wallichii* *Mesua ferrea* *Altingia excelsa* *Manlietia insignis* *Artocarpus chaplasha* *Eugenia* spp. *Tetrameles nudiflora* *Stereospermum chelonoides*
Wet	300–400	2–4	Wet forests at medium altitudes; optimum type	Closed rain forest (western form)	*Poeciloneuron indicum* *Palaquium ellipticum* *Mesua ferrea* *Dipterocarpus indicus* *Hopea parviflora* *Cullenia excelsa* *Vateria indica*
	200–300	2–3	Wet forests at medium altitudes; optimum type	(eastern form)	*Calophyllum elatum* *Dipterocarpus pilosus* *Artocarpus chaplasha* *Artocarpus heterophyllus* *Mesua ferrea* *Cinnamomum cecidodaphne* *Altingia excelsa*

Grass cover	Main types of land-use	Main crops	Livestock	State
Pseudanthistiria heteroclita *Themeda quadrivalvis* *Dichanthium annulatum* (in plains) *Arundinella pumila* *Sehima nervosum* (on the hills) *Imperata cylindrica*, *Narenga porphyrocoma*, *Saccharum spontaneum*, *Themeda arundinacea*, *Microstegium ciliatum*	Some shifting cultivation, forest utilization on scientific basis, arable cultivation with seasonal fallows, forest grazing, horticultural crops	Rice, jute, sugar cane, pulses, tobacco, mangoes bananas	Cattle, buffaloes, goats, some sheep	Maharashtra, Mysore, Orissa, Bihar, U.P. and Bengal
Arundinella–Cymbopogon *Imperata–Saccharum–Themeda* *Phragmites–Saccharum*	Shifting cultivation, intensive cultivation, short fallows, forest utilization	Jute, rice, pulses, wheat, sugar cane cinchona, rubber cashewnut (in W. Ghats), coffee (Coorg)	Cattle, buffaloes, goats	Maharashtra, Mysore, Kerala, Assam and Bengal
No grass in the undisturbed forest No grass in the undisturbed forest	Shifting cultivation, spice gardens, forest utilization for timber and forest products, plantation crops, horticultural crops	Hill rice, tapioca, cardamom, pepper, tea, rubber, bananas, jack fruits, coconuts on the coast	Cattle, buffaloes, goats, pigs, horses	Mysore, Kerala, Assam

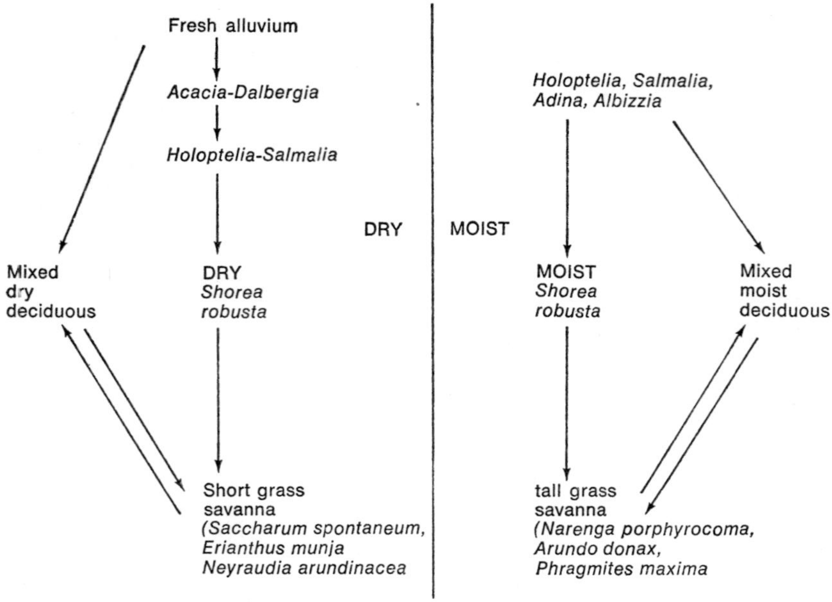

Fig. 5/5: Succession in forests of *Shorea robusta* in Uttar Pradesh, India (BHATNAGAR, 1960). The border between dry and moist forest is taken as 1,143 mm. rainfall. The xerophytic stage of dry *Shorea robusta* forest is mainly dominated by *Anogeissus latifolia*, *Acacia catechu* and *Aegle marmelos*, with *Eulaliopsis binata* most common grass.

5.8.3 ASIA

The types of grass covers and the succession to be found within them in south-east Asia and in western monsoon Asia (especially India, Pakistan and Sri Lanka) are described by WHYTE (1968a, 1972a and 1973). The outstanding example is the recognition of the sub-climax and seral types of grass cover within India, as a result of the eight-year Grassland Survey (DABADGHAO & SHANKARNARAYAN, 1972). Sequels to this have been the studies of *Iseilema*, *Sehima* and *Heteropogon* communities within the *Sehima/Dichanthium* zone (DABADGHAO & SHANKARNARAYAN, 1970); also of the grass cover types in the western Himalaya (GUPTA & NANDA, 1970). The grass covers of Maharashtra State have been classified into seven major habitat patterns and 24 minor sub-patterns, on the basis of soil, topography, hydrography and other phenological considerations (OKE, 1972).

The grass constituents in the last column headed 'vegetation' in the tabulated analyses of twelve land systems in the Nagarparkar peninsula in Pakistan (between the Tharparkar dune fields and the seasonally flooded, saline littoral plains of the Rann of Kutch) could be expressed quite differently by the grass cover ecologist (WRIGHT, 1964). It is first necessary to know the history of the present shrub/grass associations, to attempt to define the original climax covers and to analyze the subsequent effects of use and misuse over millenia of time. On that basis, it would then be possible to relate the vegetation (past and present, climax and disclimax) to geomorphology, slope and the texture and moisture-holding capacity of the soil.

BHATNAGAR (1960) shows that clear felling of moist *Shorea robusta* forests in north India tends to raise the watertable, and grass communities include *Narenga porphyrocoma*, *Arundo donax* and *Phragmites maxima*; in dry forests, pioneer grasses are *Saccharum spontaneum*, *Erianthus munja* and *Neyraudia arundinacea*. (Fig. 5/6). PANDEYA (1969) has concluded his series on the ecology of the grasslands of Sagar, Madhya Pradesh, by considering edaphic factors in the distribution of associations. SINGH & MISRA (1969) show that species diversity increases productive efficiency of the grassland ecosystem, while dominance makes the system stable, though less efficient for production, contrary to the conclusion of MCNAUGHTON (1967), that diversity on Californian grassland reduces efficiency and generates stability in the community.

5.8.4 AFRICA

RATTRAY (1960) mapped the grass covers of Africa on the basis of information available at that time, and of discussions with specialists on the vegetation of that continent. Since that date, the Land Resources Division of the British Directorate of Overseas Surveys has included the grassland resource in their overall analyses. For example, an ecological survey of Western Province, Zambia, related particularly to fodder resources (VERBOOM & BRUNT, 1970); appendix 2 to volume 2 of the report gives the main vegetation communities (based on TRAPNELL, MARTIN & ALLAN, 1950, and TRAPNELL & CLOTHIER, 1957), the grass cover types and their species composition. ASTLE (1968–9) has described the four types of grassland (virtually treeless areas) found on a ranch in Luapula Province, Zambia; because of a distinct succession of flowering in any one area, species composition appears to vary according to season of survey. HOPKINS (1968, 1970a and 1970b) also refers to seasonal changes in Nigeria. VESEY-FITZGERALD (1970) describes the origin and distribution in relation to edaphic factors of valley grasslands in East Africa.

The programme 'Afrika-Kartenwerk' of the Deutsches Forschungsgemeinschaft, initiated in 1963 following a recommendation of C. TROLL, was planned to cover the following regions (KAYSER, MANSHARD, MENSCHING & SCHULZE, 1966);

North Africa (Tunis-Sfax)
Tropical West Africa (Southern Nigeria)
Tropical East Africa (Lake Victoria)
South Africa (Lourenço-Marques – north-east Transvaal).

BADER (1967a and 1967b) discusses the problems which have arisen in connection with parts of the Lake Victoria sheet (BADER, 1974), especially the *Cymbopogon afronardus* grassland and the *Combretum/Cymbopogon* savanna.

For sub-Saharan West Africa, BOUDET & RIVIÈRE (1968) relate the tree/grass associations to the zones between specific isohyets: 150–200; 200–300; 300–400 and 400–550 mm. respectively; this begins with a ground storey of tufts of *Aristida plumosa*, *A. pungens* and *Panicum turgidum*, and reaches a steppe with *Aristida mutabilis* and *Eragrostis tremula*, giving place to a cover of tufts of *Hyparrhenia dissoluta*, *Cymbopogon giganteus* and *Andropogon gayanus*, with the annuals *Aristida mutabilis* and *Schizachyrium exile*. Species distribution within the recognized distributional zones of West Africa south of the Sahara is given as follows (BILLE, LEBRUN & RIVIÈRE, 1968/9; see also RATTRAY, 1960).

Sahelian

Andropogon gayanus, *A. pseudapricus*, *A. schirensis*, *Diheteropogon hagerupii*, *Loudetia* spp.
Hyparrhenia rufa, *Hyparrhenia* spp.
Paspalum orbicularis, *Leersia hexandra*
Echinochloa stagnina, *E. crus-pavonis*

Guinean

Hyparrhenia diplandra, *H. chrysargyrea*, *H. confinis*, *Andropogon tectorum*, *A. macrophyllus*, *Loudetia* spp., *Panicum phragmitoides*, *Brachiaria brizantha*, *Pennisetum purpureum*, *Imperata cylindrica*.

A detailed study of an area in Mali (BOUDET & ELLENBERGER, 1971) recognizes grass cover types of *Andropogon pseudapricus*, *Schizachyrium sanguineum* and *Andropogon gayanus*; four types of pasture flooded in the wet season contain respectively *Hyparrhenia rufa/Schizachyrium platyphyllum*, *Andropogon canaliculatus/Anadelphia afzeliana*, *Andropogon africanus/Sorghastrum trichopus* and *Paspalum polystachyum* or *Echinochloa stagnina*.

5.8.5 AMERICA: SOUTH

Spatial succession of grass communities in the Alto Llano of Venezuela is given by BLYDENSTEIN (1962). The main physiognomic types are:

a. wet grassland in gallery forest – *Paspalum pulchellum*, *P. chaffanjonii*, *Eragrostis acutiflora* and other humid grasses.

b. low Llano – *Mesosetum rottboelloides*, *Paspalum* spp., *Sorghastrum parviflorum*, *Andropogon selloanus*, *Axonopus purpusii*
c. palmares: *Copernicia tectorum*, *Leersia hexandra*, *Panicum laxum*
d. morichales: *Mauritia minor*, *Jussiaea lithospermifolia*, *Panicum stenodes*, *Andropogon virgatus*, *Rynchospora velutina*.
e. high Llano – *Trachypogon* savanna – *T. vestitus*, *T. montufari*, with trees, *Curatella americana*, *Byrsonima crassifolia*, *Bowdichia virgilioides*.

Five variations of *Trachypogon* savanna (e) are recognized, mainly on the basis of soil characters.

The vegetation of the white-sand savanna in Surinam comprises (a) forests and woods, (b) scrub and bushes, and (c) ground covers consisting of a single stratum (HEYLIGERS, 1963). Of the last, there are five main types of vegetation, characterized respectively by:

(i) *Bulbostylis conifera*
(ii) *Axonopus attenuatus/Lagenocarpus weigelti*
(iii) *Trachypogon plumosus*
(iv) *Rynchospora tenuis*
(v) *Lagenocarpus tremulus*

The last type has four variants, with algae or *Sphagnum antillarum* covering from 10 to 90 per cent of the surface, grasses the remainder (*Paspalum pulchellum*, *Panicum micranthum*, *P. polycomum*, *P. nervosum*).

5.9 **Succession in time and place**

5.9.1 Categories of Grass Covers

After allowing for a continuing state of imbalance and heterogeneity in grass covers, it is nevertheless possible to recognize stages in regression or progression on any given site (PANDEYA, 1961). The basic environmental factors that govern the composition of a grass community are soil, climate, exposure, altitude and slope. In addition to these relatively permanent factors, there are others which are in a constant state of change in nature and intensity. Under their influence, some species become rare or absent, and new ones more adapted to the new conditions take their place. It is the object of the practising ecologist to be able to recognize these seral stages, to diagnose their origins, and to indicate what seral stages might be expected to appear if the governing factors were reduced or increased in intensity.

Although the overall picture is still patchy due to lack of sufficient analyses over the whole intertropical zone, some preliminary indicators are already appearing. There are certain species which are usually found

Table 5/2

INDIA: Succession in types of grass covers

(a)

Sehima/Dichanthium

		↑ ↓		
Eroded slopes	(Gravelly soils)		(Well-developed soils)	↕ Increasing soil moisture
Burning	Protection		Grazing	*Iseilema*
Cymbopogon	*Themeda*			↕
Pseudanthistiria				*Ischaemum indicum*
	Chrysopogon		*Bothriochloa*	
	Protection		Grazing	
	Heteropogon		*Eremopogon*	
	Protection		Grazing	

Aristida/Eragrostis/Melanocenchris

(b)

Dichanthium/Cenchrus/Lasiurus

	↑ ↓	
Protection		Grazing
		Cenchrus/Lasiurus
Protection		Grazing
		Cynodon/Eleusine
Sporobolus marginatus		Grazing
Protection		*Aristida*
Chloris		*Cenchrus biflorus*
(Compact soil)		(Loose soil)

(c)

Phragmites/Saccharum/Imperata (*Savanna*)

↑ Protection ↑ | ↓ Burning and cutting ↓ | ↕ Grazing ↕ *Vetiveria*

Saccharum/Imperata/Sclerostachya

↑ Protection ↑ | ↓ Burning and grazing ↓

Desmostachya/Imperata (depauperate)

↑ ↓

Sporobolus/Paspalum/Chrysopogon

(d)

Themeda/Arundinella

↑ Protection ↑ | ↓ Grazing ↓ | ↓ Burning ↓ *Dimeria*

Arundinella/Chrysopogon

↑ Protection ↑ | ↓ Grazing ↓

Heteropogon/Bothriochloa

↑ Protection ↖ | ↓ Grazing ↙

Cynodon

Table 5/2 (continued)

(e) Temperate Alpine Type (Himalaya)

The grass vegetation of this primarily non-monsoonal cover is composed of the following species:

Agropyron canaliculatum	*Chrysopogon gryllus*
Agrostis canina	*Dactylis glomerata*
A. filipes	*Danthonia jacquemontii*
A. munroana	*Koeleria cristata*
A. myriantha	*Phleum alpinum*
Andropogon tristis	*Poa pratensis*
Calamagrostis epigejos	*Stipa concinna*

Associated perennial species with a contribution of more than 10 per cent in the composition are:

Agrostis pilosula	*Festuca valesiaca*
Brachypodium sylvaticum	*Helictotrichon asperum*
Bromus ramosus	*Muhlenbergia* sp.
Calamagrostis emodensis	*Poa alpina*
Eragrostis nigra	*Trisetum* sp.
Festuca lucida	

(f) Nilgiri grasslands

Andropogon polyptychus/Eulalia
↕
Eulalia/Themeda/Ischaemum
↕
Chrysopogon/Ischaemum/Themeda
↕
Ischaemum/Chrysopogon
↕
Ischaemum/Tripogon
↕
Tripogon/Eragrostis/Arundinella

Source: (a) to (e) – P. M. Dabadghao, Grassland Survey of India.
(f) – S. C. Gupta, S. Chinnamani & N. D. Rege, Personal communication

high in the succession, in the absence of intensive operation of use factors. As grazing intensity increases, the more palatable species will be reduced in percentage cover; the bunch grasses of the climax will be damaged because of their slow growth habit, which becomes exposed to defoliation at an earlier, more critical stage than the other, more rapidly growing perennials in the same community. Then there is a group of species that always appears to belong to lower stages in succession, tough and less palatable and adapted to greater soil aridity. These are the pyrophytes that are resistant to fire, or may even be stimulated by it. Finally, there are the annuals representing the last-ditch attempt of a genus to perpetuate itself in conditions in which its perennial ancestors can no longer persist. A categorization of the major species of intertropical grass covers in this way would be of great value, in both taxonomic geography and taxonomy. Such a grouping may be seen as a background theme through these Chapters.

5.9.2 Asia

The study of stages in regression within the major grass covers on different soils and under different management influences was an essential part of the Grassland Survey of India (Table 5/2). It is to be hoped that some of the surveyors may be able to return to selected sites from time to time to record any trend that may become apparent. It is becoming increasingly obvious that botanists and ecologists everywhere have little idea what their vegetations and grass covers were like a hundred, fifty or even twenty-five years ago.

Verboom (1968) has studied grass associations and succession in Pahang, Malaysia, in a tropical rain forest area of the type discussed in Chapter 2.4. Seven distinct grass covers are recognized on the basis of place of occurrence. These are among the secondary successions which would ultimately revert to forest. The shade-tolerant species noted in 2.4 and 9.5.2. are actually forest fringe grasses. The forest clearings have the usual pioneer associations – *Paspalum conjugatum*, *Cynodon dactylon*, *Digitaria* spp., *Echinochloa colonum*, *Eleusine indica*, etc. Species of wet padang and roadsides are listed. See Fig. 5/6.

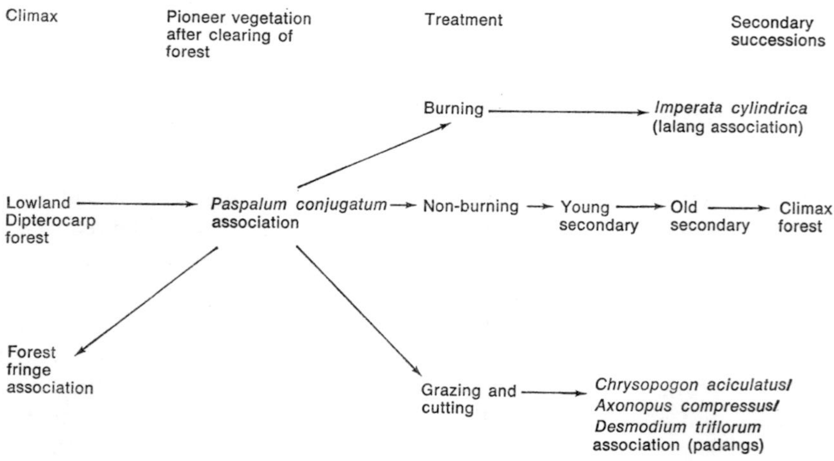

Fig. 5/6: Relation between land management and plant succession in Malaysia (Verboom, 1968).

5.9.3 Africa

A dynamic approach to the survey and planned utilization of grazing land has been adopted by French workers in West Africa and Madagascar (Piot, 1966; Valenza, 1970; Granier, 1965; Garnier, Lahore & Dubois, 1968). See section 5.8.4 for examples of spatial succession in West Africa.

Lansbury, Rose-Innes & Mabay (1965) note that the 'grasslands' of Ghana are mostly tree or shrub savannas or savanna woodlands, as defined at Yangambi (C.S.A., 1956). They consist of a continuous grassy ground cover, interspersed with trees and shrubs of variable height and density; the grass covers are seral in character and proclimax in status, progression to a predominantly woody climax being prevented by frequent fires and other disturbances.

Dominant Species

Sandy soils	Clay soils
Schizachyrium schweinfurthii	*Vetiveria fulvibarbis*
Andropogon canaliculatus	*Andropogon canaliculatus*
Vetiveria fulvibarbis	

Jacques-Félix (1968) returned to an area in the Cameroons that he had visited in 1939. At that time he had noted vestiges of a forest of *Khaya grandifolia*; now the region has 'une allure très naturelle' and one would never suspect the earlier forest climax. On certain plateaux, *Urelytrum fasciculatum* was the dominant grass; now the development of animal husbandry has led to its replacement by *Sporobolus pyramidalis*, with rare stands of *Brachiaria brizantha*.

In Somalia, Glover (1947) and other British officers reported the presence of considerable areas of perennial grass; now they are absent (Box, 1968) or present only in relict areas, to indicate the potential (*Themeda triandra*, *Chrysopogon aucheri*, *Bothriochloa insculpta*, *Hyparrhenia hirta* and other tall grasses). In the high plateau country (the Haud), grazing has been less destructive, and *Cenchrus ciliaris*, *Sporobolus* sp. and, near water, dense stands of *Aristida kelleri* are found.

Owen & Brzostowski (1966) have discussed the possibilities of ranching development on the Kongwa Plain, Tanzania, on the basis of the proportions of annual and perennial species in the grass cover, and the tendency for progression from a cover of annual grasses and herbs to perennial shrubs and to taller bush species and the common components of bush thickets. In those parts of Uganda (Buganda and Busoga bordering Lake Victoria), there are a number of ecological vegetation zones in an area occupied by twenty-two dairy farms, but pressure of cultivation has left little more than relict areas of the climax. Reference is made to the possible climax communities (Langdale-Brown, Osmaston & Wilson, 1964), in order to assess the true dairy potential (THORNTON, LONG & MARSHALL, 1969; LONG, THORNTON, NDYANABO, MARSHALL & SSEKAB, 1970; see also THORNTON & MARSHALL, 1971). Nutritive value in relation to specific composition and succession in grass covers has been studied in Ankole and the Queen Elizabeth National Park, Uganda (THORNTON, LONG & MARSHALL, 1968; LONG, THORNTON & MARSHALL, 1969; LONG, NDYANABO, MARSHALL & THORNTON, 1969). A similar study of nutritive value of seral stages and a good perennial grass cover (*Cynodon dactylon*, *Eragrostis superba*, *Digitaria scalarum* and *Cymbopogon* sp.) has been made in the browse and grass range pastures of the semi-arid upland area of Kenya – 1,575 m. elevation in Rift Valley (McKAY & FRANDSEN, 1969).

The British Land Resources Division (VERBOOM & BRUNT, 1970) notes in Western Zambia the change in the composition of the open grassland, compared with the grass cover found in woodland under a fire regime. The tendency for many pastures to revert to a *Sporobolus*-dominated cover when overgrazed has been noted. On Lake-dune Barotse sands and Lake Basin Soils, where grassland often has a high proportion of *Brachiaria dura*, overgrazing may lead to dominance of relatively unpalatable *Aristida* species; if the shrub growth of *Bauhinia*

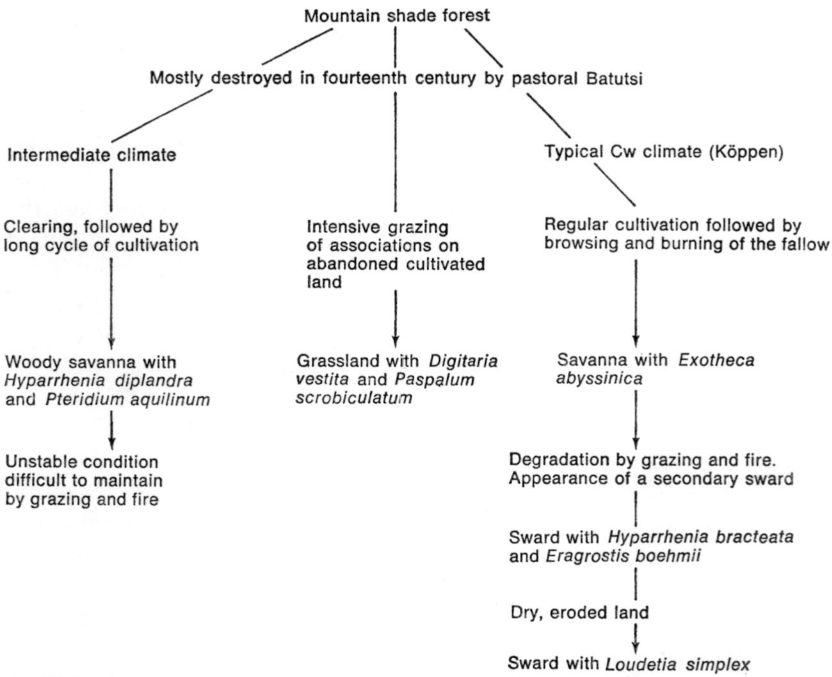

Fig. 5/7: The floristic evolution of a permanent grass cover of *Exotheca abyssinica* and *Eragrostis boehmii* in Rwanda and Burundi at altitudes of 1,800 to 2,500 m. (COMPÈRE, 1968).

mucronata and *Baphia obovata* is not slashed, it will slowly increase and shade out much of the grass cover. In low-lying areas, succession and hence composition are functions of depth of flooding; the most nutritious sward may be maintained by controlling depth of flooding and level of watertable. VERBOOM (1966) finds that the grass covers of Barotseland differ so much from the remainder of Zambia that one may regard it as a special environment with affinities with south-east Angola, the Caprivi Strip and north Botswana.

Succession in a high-altitude tropical 'grassland' has been studied by COMPÈRE (1968) in Rwanda and Burundi (Fig. 5/7). BILLE (1965) has studied succession in natural pastures of the high plateaux of the Central African Republic (Tables 5/3 and 5/4).

Aspect and altitude are important factors in determining the extent of the four vegetation types in the Upper Mgeni Catchment, 25 km. north-west of Pietermaritzburg (MOLL, 1968):

Themeda/Hyparrhenia grassland (below 1,069 m.)

Moist transitional *Themeda/Hyparrhenia* grassland (914 to 1,219 m.)

Themeda/Aristida grassland and Mistbelt Mixed *Podocarpus* Forest (1,219 to 1,372 m.)

Themeda/Trachypogon grassland and Mountain *Podocarpus* Forest (above 1,372 m.)

The vegetation history of the Catchment is related to fire and land use. Fire in early spring was a natural factor, but autumn burning followed human settlement and the grassland changed from *Themeda* to *Aristida*. Land use changed from extensive to semi-intensive with the coming of the Europeans; their domestic stock were selective graziers, and eliminated the palatable *Themeda* in favour of the unpalatable *Aristida*.

5.9.4 AMERICA: CARIBBEAN ISLANDS

The cleared pasture lands of the Caribbean region may be grouped into ten main classes on the basis of soil moisture and topography. Twenty six grass communities (eleven major and fifteen minor) have been recognized in Puerto Rico by MOLINARI (1952). By studies of dominant, principal and secondary species, the types have been correlated with soil groups and types of rainfall. Most are also found in Santo Domingo, Cuba, Haiti and Jamaica; some occur in the Virgin Islands and Lesser Antilles (MOLINARI, *op. cit.*). *Panicum purpurascens* occurs in moist lowlands throughout, *Stenotaphrum secundatum* along moist limestone coastal plains on the windward sides; *Cynodon dactylon*, *Sporobolus virginicus* and *Chloris inflata* are the dominant pastures of the dune formations and pure sandy soils fringing the ocean; *Sporobolus*

Table 5/3

Central African Republic: Grass succession in regions of semi-permanent occupation, classing the undergrazed and overgrazed pastures on a scale 1–11 (Bille 1965).

	1	2	3	4	5	6	7	8	9	10	11
Andropogon tectorum	+										
Hyparrhenia diplandra	+	+									
Panicum phragmitoides	+	+	+	+	+						
Hyparrhenia chrysargyrea		+	+	+							
Hyparrhenia rufa		+	+								
Urelytrum thyrsioides			+	+	+						
Brachiaria brizantha				+	+	+	+				
Paspalum commersonii											
Setaria sphacelata					+	+	+	+			
Eleusine indica											
Dactyloctenium aegyptiacum							+	+			
Eragrostis spp.											
Sporobolus spp.							+	+	+		
Adventices + sol nu								+	+	+	+
Solanum spp.											+

indicus, *Axonopus compressus* and *Andropogon* spp. occur on the moist mountainous regions.

In a coarse type of grassland among characteristic trees on true laterite soils of serpentine origin in Puerto Rico and Cuba, the dominant grasses are *Leptocoryphium lanatum* and *Andropogon semiberbis*.

Table 5/4

Central African Republic: Succession on pastures on laterite soils grazed by migratory stock, classing undergrazed and overgrazed pastures on a scale 1–8 (Bille 1965).

	1	2	3	4	5	6	7	8
Hyparrhenia dissoluta	+							
Hyparrhenia filipendula	+							
Andropogon gabonensis	+	+						
Hyparrhenia gracilescens	+	+						
Hyparrhenia soluta	+	+	+	+				
Andropogon schirensis		+						
Hyparrhenia notolasia		+	+	+				
Loudetia arundinacea		+	+	+	+	+		
Panicum phragmitoides			+	+	+	+		
Loudetia kagerensis				+	+			
Hyparrhenia rufa				+	+			
Hyparrhenia chrysargyrea				+	+	+		
Eragrostis – Sporobolus						+	+	
Adventices + Sol nu							+	+
Solanum distichum								+

Bothriochloa pertusa (OAKES, 1968) and *Andropogon intermedium* var. *acidula* are important in the Virgin Islands but not in Puerto Rico or other Greater Antilles. *Andropogon* (= *Dichanthium*?) *caricosus* forms extensive pastures in Cuba, but is sporadic in Puerto Rico.

5.10 AMERICA: CARIBBEAN ISLANDS: Succession on abandoned cultivated land

Most of the present native pasture lands in Puerto Rico has been under crops at one time or another; some areas have been undisturbed for 40 years, others for a few months to several years (MOLINARI, 1949). In most upland areas of the north interior mountain region, trees start reversion to the forest stage; the grasses of the open parkland disappear and are replaced under the forest canopy by shade-tolerant grasses, *Lasiacis divaricata, L. ruscifolia, Pharus glaber, Ichnanthus pallens, Olyra latifolia, Oplismenus setarius* and others.

Studies of succession show the role of annuals as first pioneers and as last relicts under adverse conditions, and the relative place in succession of perennial species.

Northern Coastal Plain Grassland

1st Stage: Annuals occupying land for only a short period – *Echinochloa colonum, Eriochloa punctata, Digitaria horizontalis, Eleusine indica, Cenchrus echinatus, Paspalum fimbriatum.*

2nd Stage: Long-lived annuals forming dense sod – *Digitaria horizontalis, D. sanguinalis*; scattered seedlings of perennials – *Paspalum conjugatum, P. notatum, Axonopus compressus, Sporobolus indicus, Stenotaphrum secundatum*; first perennials following annuals in heavily grazed areas – *Paspalum conjugatum, Sporobolus indicus.*

3rd Stage: These last species are dominant at this stage; being less palatable than the annuals, they grow undisturbed for one to three years after abandonment. Ultimately, depending on soil and climatic factors and intensity of grazing, the grasses are: *Axonopus compressus, Paspalum notatum, Stenotaphrum secundatum* and *Sporobolus indicus.*

Other successions are recorded for:

a. the abandoned cotton lands of the Playa Plains, sandy lands near the seashore:

1st Stage: *Panicum adspersum, Eragrostis ciliaris*

2nd Stage: *Dactyloctenium aegyptiacum, Digitaria horizontalis, D. sanguinalis*

and on sand dunes: dominant *Sporobolus virginicus* with *Philoxerus vermicularis* and some *Cenchrus pauciflorus*, *C. myosuroides*, *Chloris inflata* and *Tricholaena repens*;

b. the fresh water lagoons and swamps, in the submerged, floating, reed-swamp and sedge-meadow stage, the last composed of many species of sedges with *Paspalum distichum*, *Sporobolus virginicus*, *Sacciolepis striata*, *Leersia hexandra* and *Paspalum millegranum*; and

c. the dry grasslands of the south and south-west:

1st Stage: *Panicum reptans*, *P. adspersum*, *Dactyloctenium aegyptiacum*, *Digitaria horizontalis*, *Eleusine indica*, *Paspalum fimbriatum* and *Cenchrus echinatus*

2nd Stage: *Chloris inflata*, *Cynodon dactylon* leading to a climax of *Bouteloua heterostega* and *Sporobolus indicus*.

6 ECOLOGICAL MANAGEMENT OF NATURAL RESOURCES

6.1 Science and practice of management

The literature on the management of natural grazing lands on the basis of the principles of plant ecology, and within the limitations imposed by the need to conserve soil and water and to prevent excessive desiccation of the environment, generally comes under the broad heading of range management. Mostly of American origin or written by American specialists working in other lands, it is kept at an eminently practical level for its public of graziers and animal husbandmen. Since this literature deals with the management of livestock on natural vegetation, it is a little difficult to extract those principles and techniques of interest to the vegetation scientist.

PETERSON (1969) presented an FAO analysis of pastoral problems to an African audience concerned with grassland production in Africa south of the Sahara: 'In the identification of grassland problems, it is necessary to see grasslands in the context of the whole. This helps to assure that the most appropriate and most relevant action is taken. The most important major inter-connected components to consider in an analysis are: 1. the biological system, 2. management, and 3. social and economic influences. In the biological system, the natural limitations imposed by environments, plants and animals, and their interactions, should be identified, in order to know the type of man-supplied inputs which are needed. Some type of management is always applied to the biological system. Mostly in Africa it is traditional and designed for little more than subsistence .... Social and economic influences may impose strong constraints to change .... Of particular importance is the discovery of basic incentives which will encourage the grazier to want to sell more animals of higher quality .... A major thrust must be directed at improving the output from native grasslands and their critical supplementation through forage production on arable lands'.

6.2 Levels of production

In such developments, one may consider four production levels, involving progressive increases in capital expenditure to provide a more reliable seasonal availability of better quality plant material, leading through the fodder production and utilization chain to the end objective, production of specific requirements from an efficient livestock industry:

Production levels:

primary:	using the resource as it stands, by adjusting stock pressures and movements, adoption of sylvo-pastoral technique, selective opening and closing of watering places;
secondary:	improving the ground cover resource by removing competitive trees and shrubs;
tertiary:	introducing foreign species of grasses, and especially legumes, into the existing resource; and
quaternary:	removing the natural resource and replacing it with artificial or synthetic plant communities for use as pasturage or for fodder production (Chapter 8).

6.3 Management on basis of condition analysis

Humphrey (1966) introduced the botanical and ecological components of the technique of range condition analysis adopted in the United States of America to an audience of tropical specialists at the Ninth International Grassland Congress in Brazil. The following paraphrased summary may suffice in the present context (see 4.7).

The ecology of a region must be thoroughly known before making range condition analyses on which management practices may be founded. However, a range condition analysis is not a purely ecological classification. It is rather a practical means of rating ranges to determine whether, under the current type of management, they are producing the crop of forage that they should. Five condition classes may be used: excellent, good, fair, poor, and very poor or depleted. To recognize an excellent condition, one must know the ecology of the area, its vegetation and the indicator value and forage value of its component species; also the climate, the production potential of the soils, and their genesis. One must assess the history of the vegetation in relation to past climatic fluctuations and crises, to use and misuse.

The five criteria of condition to be adopted by the range research worker and adviser have been noted in 4.7: a. forage composition; b. ground cover; c. plant vigour; d. litter; e. erosion, with a possible sixth, reproduction.

The vegetation or forage types are usually separated on a life-form basis, e.g. forest, woodland, savanna, chaparral, desert scrub, grassland. Each type may, if necessary, be further subdivided on a floristic basis. This initial mapping into vegetation types is necessary because each occupies its own ecological niche. Condition classes are delimited within each vegetation type.

Trend in range condition is also important in assessing whether present management should be changed, whether the range cover is improving, deteriorating or remaining static. Trend may be determined by two methods. The ideal would be a combination of the two.

The interim method records the vegetation growing on an area at a given time, records it again in a later year, and analyzes the changes. This gives a permanent record but does involve at least two surveys. It also assumes that the analyst knows the successional ecology of his vegetation types. 'There is a general human tendency when working with noncultivated species to assume a high degree of permanence to the plants one encounters under natural conditions' (HUMPHREY, *op. cit.*).

The current appraisal method observes and evaluates certain characteristics of the soil and vegetation, whether erosion is increasing, static or decreasing, and what plants are maintaining themselves and reproducing. The general objective of management on range lands should be maximum forage production for a particular kind of animal, domesticated or wild, at a particular time of the year or for the full year. A range is analyzed as to present condition and direction of trend as a kind of diagnosis to assist in developing effective plans for management, leading to maximum production of livestock products in the legion of local ecosystems that exist, for example, throughout the intertropical zone.

6.4 The ecological component in range management

A review of the ecological bases for the management of intertropical grazing lands would call for a detailed study of an extensive and scattered literature of a scientific and practical nature. It would be necessary to compare the indigenous systems which have evolved over the years in many different ecoclimates and in many different seral stages in vegetation succession.

The East African Range Classification Committee (PRATT, GREENWAY & GWYNNE, 1966) considers that the arid zone of northern Kenya comprises the semi-desert together with contiguous lands of higher rainfall, nearly one half of the total area of Kenya (East African Royal Commission, 1955). Mt. Kulal (2290 m.) and Mt. Marsabit (1430 m.) both carry forest at their upper elevations and are flanked by a ring of semi-arid country, characterized by woodland species of *Acacia* and perennial grasses such as *Themeda triandra* (PRATT, 1968, 1969). The vegetation of the basement plains is dwarf-shrub grassland; an open scatter of larger shrubs including *Acacia reficiens* ssp. *misera*, with dwarf shrubs, species of *Indigofera* (especially *I. spinosa*), *Sericocomopsis*, *Barleria* and *Disperma*. The commonest grasses are annual species of *Aristida* and *A. papposa*, but there is much variation; in some areas *Cenchrus pennisetiformis* may be dominant, or even the perennial species of *Dactyloctenium*, *Latipes senegalensis*, and species of *Enneapogon* normally classed as perennials.

PRATT (*op. cit.*) finds that the traditional systems of management of the Hamitic or Nilo-Hamitic pastoralists are sound and no radical change is proposed. The present grazing units are the tribal areas, and these are often too large for development purposes. Research on range management at the East African Agricultural and Forestry Research Organization has been described by McKAY (1970); by ADEGBOLA, ONAYINKA & EWEJE (1968) for Nigeria; FAO (1970) for Malawi; GARNIER, LAHORE & DUBOIS (1968) for Madagascar.

In Rwanda and Burundi, within the pattern of succession at altitudes of 1,800 to 2,500 m. shown in Fig. 5/7 the maintenance of a favourable floristic equilibrium in which *Exotheca abyssinica* and *Hyparrhenia bracteata* are dominant is favoured by the subdivision of the pastures into eight equal plots, and by the adoption of a grazing plan ensuring a rest period of 30 to 40 days (COMPÈRE, 1968). The objects of various types of rest have been defined for the South African grass veld (TAINTON, 1971):

(i) seeding rests
(ii) rests for seedling establishment
(iii) rests for tillering
(iv) rests for herbage accumulation, for the accumulation of palatable herbage, or for burning for the control of encroaching bush
(v) rests for restoration of vigour, and
(vi) short-term withdrawal of little direct benefit to veld.

TAINTON, BOOYSEN & SCOTT (1970) review the literature on the effect of cutting height on the production of some twenty to twenty-five temperate and subtropical species of grass. In their own studies in which an objective was the maintenance of a correct proportion between *Themeda triandra* and *Tristachya hispida*, it was found that herbage yields declined with increased height of cutting.

'High cutting induced tiller development at higher nodes than did low cutting, resulting in a lack of contact between the elevated tiller bases and the soil and in the elimination of differences between the proportions of each such tiller harvested and left by cutting to different heights.'

Over a period of eleven years, two veld types (climax *Themeda* and pioneer *Eragrostis/Aristida*) have responded to fertilization with nitrogen and phosphorus (VORSTER & MOSTERT, 1968).

'The best response occurred during years with high rainfall. In general, nitrogen fertilization increased the hay, protein and phosphorus production of *Themeda* veld, but caused a successional retrogression in the plant cover. The effect of nitrogenous fertilization on the *Eragrostis/*

Aristida veld was much more marked than on the *Themeda* ve
of plant succession being positive. Phosphorus fertilization
little influence except in interaction with nitrogen, where
yield.'

The botanical and ecological bases of the managem
lands in India have been fully described (WHYTE, 1968a ,..
in India is governed by socio-economic/religious factors that preve..
the evolution of a scientific system of management based upon the full knowledge of grass succession which is now available. KHAN (1970, 1971a ,1971b) has described the current situation in Pakistan, for which JOHNSTON & HUSSAIN (1963) have already proposed grass covers comparable to those in western India. In both countries will be seen an increasing tendency to polarization in grassland and fodder development (1.5 and WHYTE, 1971).

This section ends with the conclusions of WILLIAMS (1968) on certain ecological concepts that may appear helpful in studies of grazing as a manipulation:

1. 'substantial changes in flora and fauna can be anticipated with grazing only if the plant community is near the original state;
2. on disclimax communities, heavy and continuous grazing pressure will be required before there are significant shifts in perennial dominance, and unusual manipulations will be needed before major shifts in botanical composition can occur, assuming that no species invade from outside the community;
3. sown pasture consisting of free-seeding annuals, and possibly the grass *Phalaris tuberosa*, could be considered as disclimax communities;
4. the domestic herbivore is in the process of constructing or modifying both plant and animal communities at the present time;
5. the selection of the physiological factors involved in success or failure should be made with due regard for the ecological status of the species under investigation, i.e. the historical antecedents of the present Australian fauna and flora should not be neglected' (2.2).

6.5 **Prevention of progression to forest**

Perhaps it is particularly in Africa that those responsible for ecological management of natural grazing lands are faced with the problem of bush encroachment, which is an expression of a natural ecological tendency of the vegetation to progress upwards through successive stages of shrubs, bush and small trees to the local woodland or forest climax. Burning is the main tool adopted to control such growth, and also to remove old, dead unpalatable herbage that smothers new growth of the grasses.

A discussion whether bush is to be regarded as an infestation or an asset depends on the type of livestock being produced – cattle, goats or wildlife (Ivens, 1972). Some argue that cattle obtain nourishment by browsing, especially during dry seasons, others that carrying capacity of grass from which bush has been cleared is doubled or trebled. A certain proportion of bushy growth may be retained for shade.

Two reviews discuss the problem of 'to burn or not to burn' in general terms, in relation to West Africa and South Africa respectively (3.1.5). Ramsay & Rose-Innes (1963) report on some quantitative observations on the effects of fire on the Guinea savanna vegetation of northern Ghana over a period of eleven years. They suggest widespread planned experimentation on similar lines, an international effort on a continental scale, at intervals of about 300 to 400 miles, along one or more transects stretching from the southern limits of the savanna country northwards to the Sahara.

'Whatever the merits or demerits of grass burning may be, the fact remains that, at the present time and in present socio-economic conditions, fire is the only effective tool available to the peasant farmer for clearing and keeping land free of encroaching woody growth, for producing a uniform grass cover free of harsh, unpalatable stubble at each new grazing season, and for promoting a quick flush of off-season green growth for his animals. The practice will undoubtedly continue for a long time to come and the immediate problem is how to use it to the best advantage . . . interests which have to be catered for are chiefly those of forestry, agriculture and soil conservation.'

To the north of the proposed transect, French workers make frequent reference to the use of fire, for example Peyre de Fabrègues (1965) refers to running fires in the traditional management of pastures in the Republic of Niger; such a fire at the beginning of the dry season on an annual-dominant cover destroys the reserve of feed for the year.

The second review is that by Scott (1970) who records that veld fires have been a feature of the African landscape since time immemorial (3.1.5). In terms of ecological management, when fire has been prevented, plant succession has resulted in bush encroachment in the savanna or in fynbos encroachment in others (fynbos next stage in succession above grassland in potential forest areas, and quite useless for grazing). Fire results in much old grass being wasted, but if this is not removed, grazing deteriorates. Old grass may be removed by fire, or mowing, or by the grazing of animals fed high protein supplements. All methods other than mowing may be equally deleterious.

Other papers from Africa include Thomas & Pratt (1967) on effects of controlled burning of secondary thicket in upland *Acacia* woodland in the drier areas of Kenya; Pratt & Knight (1971) on effects of controlled burning and grazing management on *Tarchonanthus/Acacia* thicket; PRATT (1971) on the effects of fenuron pellets in which a correlation is reported between quantity of combustile grass, fire temperature and percentage kill-subsequent burning may have general application in the use of arboricides; ROSS & HARRINGTON (1969) on the practical aspects of implementing a controlled burning scheme in a national game park in Uganda. HARRINGTON (1973) has studied the fire ecology of the grasslands of Ankole, Uganda, where the vegetation (LANGDALE-BROWN, OSMASTON & WILSON, 1964) is dry *Acacia* savanna with the herb layer dominated by *Themeda triandra* in the valleys and by *Cymbopogon afronardus* on the hills. The implications for range management are that pasture on which *C. afronardus* is dominant needs to be burnt when it becomes choked with litter, but the burn should not take place until the rains have started. Pasture free of *C. afronardus* should be controlled by the grazing animal; burning encourages flowering and reduces the quantity of herbage.

In Australia, the views of TOTHILL (1971a, 1971b) on the burning of vegetation in Queensland, with reference to *Heteropogon contortus*, have been summarized in Rural Research in C.S.I.R.O., No. 74, 1971, and appropriate advice is given to farmers. Five factors favour dominance of this species in annually burnt pastures in Queensland.

- of primary importance, the seed becomes buried in the top half inch of the soil, gaining protection from fire and the elements;
- fire removes ground cover – exposing the soil to increased solar radiation which raises its daily maximum temperature;
- given adequate soil moisture, this can bring about germination when temperatures are otherwise too low;
- fire destroys some of the weaker perennial plants – reducing the level of potential competition from well-established plants;
- fire destroys much of the litter and seed at the soil surface – minimizing the potential competition from seeds that may otherwise germinate.

NORMAN (1969) has reported on the effect of burning and seasonal rainfall on native pasture at Katherine, Northern Territory, and PERRY (1970) on the effects on grass and browse production of the felling by various methods of *Acacia aneura* with a ground cover of *Aristida contorta* (more than half the field), *Digitaria brownei*, *Sida cardiophylla*, *Enneapogon polyphyllus*, *Bassia costata* and *Psilotus helipteroides*.

While recognizing the dominant role of fire in the evolution of tropical American grasslands and in their maintenance in their present form, BLYDENSTEIN (1968) states that:

'Precious little information exists on the effect of fire on the tropical vegetation of Latin America .... With the importance of fire in the ecological balance of these grasslands, the control of fire, as opposed to its elimination, should be a basic subject for investigation. We know too little about the effect of season or interval of burning on forage production, brush control, soil conditions and all the other factors which contribute to the ecological balance in producing grasslands. The studies done in other environments, with different species, provide an excellent guide as to the methods and designs which could be adopted to carry out the studies, but the applicable knowledge must come from local studies.'

6.6 Forestry working plan

At least one State of the Indian Union, Gujarat, has prepared working plans for the management of areas of grassland for which the State Forest Department is responsible (PATEL, 1969). Plans for the grassland (vidis) in Saurashtra and in Panchmahals, with a special provision for the Junagadh Division, have recently been revised. In the earlier plan for the Junagadh Division, it was the custom for the vidis to be handed over to the grass section of the Forest Department at the beginning of the monsoon, for grass growth and collection. The vidis were then returned to the Forest Department each year for exploitation of the trees on a ten-year cycle. This dual control caused unpalatable grasses and useless shrubs to increase, and the yield and quality of the fodder to decrease. In order to obviate this, cultural operations are recommended for both the trees and the grass covers.

Grass cutting should start in September and collection and despatch of the hay completed by end-January. The vidis should be closed for grazing from 1 June to 31 March; if absolutely necessary, they may be opened from 1 April, but for cows only, at a rate of one beast per 4 hectares. Ten per cent of the area of the annual coupe is to be regenerated artificially with fruit, fodder and shade trees, especially on hilltops, eroding places, blanks and patches dominated by shrubs. In the grass areas, all obnoxious herbs and thorny bushes are cut back or removed, inferior grasses cut before seeding, superior grasses permitted to seed, and the whole area oversown with broadcast seed in early June.

6.7 Shrubs and trees in tropical grazing lands

Although the main emphasis in this book is on the herbaceous members of the Gramineae and Leguminosae, the woody plants that are browsed by wild and domestic livestock must not be overlooked. The whole subject of the biology and utilization of wildland shrubs was considered at an international symposium at Utah State University, July 1971 (MCKELL, BLAISDELL & GOODIN, 1972).

'Shrubs range from some of the highest mountain elevations to the lowest. They extend from the foothills out into the drier desert areas where most grasses fail to accompany them. Only the ephemeral opportunistic annual grasses and forbs are the associated species for many shrub communities in desert and saline areas. Even under such conditions, shrubs offer certain advantages because of their productivity, palatability, nutritional qualities, value as wildlife habitat, cover for the soil, and general role in ecosystem functioning.'

Geographical reviews relate to the Indian subcontinent (K. M. M. DAKSHINI), Australia (S. L. EVERIST), Asia (M. P. PETROV), South America (A. SORIANO), and North America, including Mexican tropics (W. G. MCGINNIES). Other papers relevant to the present context are: browse and cover for wildlife (W. L. ROBINETTE); evolution and diversity of arid-land shrubs (G. L. STEBBINS); genetic improvement in crop species as contrasted with possibilities in shrubs (H. C. STUTZ); salt desert shrub response to grazing use (R. C. HOLMGREN & S. S. HUTCHINGS); the role of shrubs in nutrient cycling (J. L. CHARLEY); papers on physiological and salinity stress, gas exchange, carbohydrates; nutritive value and animal use; regeneration; future of shrubs in arid lands.

6.8 The sylvo-pastoral system

'Browsing preceded grazing by a good many millions of years' (GRAY, 1970). While in some parts of the intertropical belt, the primary objective is to provide a pure grass sward, in others a balance between tree and shrub cover and grass ground cover is recommended. The latter arises more particularly in the former French territories in sub-Saharan West Africa, where the tree/shrub/grass communities are found (5.2). Under these conditions it becomes debatable whether the use of the land is the responsibility of the forester and soil conservationist or the specialist in grazing resources. In Cameroon, PIOT (1969, 1970) has studied communities under the widely different conditions obtaining on granitic and basaltic soils. Practical considerations relate to the possibilities of complementary feeding, the correct treatment of the woody species and the increasing necessity to keep animals permanently on the grazing lands. BARTTHA (1970) discusses the specific ecological characteristics of the Sahelian zone, and the nutrition of the nomadic herds based on bush, shrub and herbaceous vegetation.

In regions of West Africa where nomadic grazing ensures maximum utilization of all types of forage available, seasonal use of the woody plants on the cultivated land is adopted since they stay green longer. But with a doubling of the human population every 25 years, and the development of waste land for arable cropping and the swamps for rice, even this resource will become exhausted. It may be desirable to plant edible shrubs such as *Acacia albida* and *Pterocarpus erinaceus* as windbreaks, for lopping during the dry season (BOUDET, 1970).

In the Northern Province of Zambia, the Research Branch of the Ministry of Rural Development (ZAMBIA, 1970) has studied the management of Miombo woodland. The system adopted utilizes resources to the maximum, providing for increased production of grass and of protein from browse. It appears to be relatively cheap and well suited to the local peoples, who follow similar clearing methods in the Chitemene system. Further work is necessary to achieve a qualitative balance between ground herbage and browse on the trees, and to improve the sward further by introducing perennial legumes.

Of the small tree and shrub vegetation of western Rajasthan, *Zizyphus nummularia*, *Prosopis spicigera*, *Calligonum polygonoides*, *Capparis aphylla*, *C. decidua* and *Balanites roxburghii* represent valuable sources of fodder, the first-named being by far the most important (KAUL & GANGULI, 1963). The leaves of *Zizyphus* are rich in digestible crude protein and minerals; leaves are still abundant when the associated grasses on these depleted ranges have dried. The shrubs provide a microhabitat in which better perennial grasses can grow, and from which they can spread under favourable management. By ocular estimate, it is found that fully, medium and poorly stocked stands of this shrub correspond to 18, 14 and 11 per cent density respectively.

7 ADDITION OF LEGUMES TO SWARD

Species of tropical legumes may be added to grass-dominated swards under certain climatic conditions, by overseeding with minimal cultivation to improve the growth and yield of the accompanying grasses (HENZELL & NORRIS, 1962), and to raise the protein content of the animals' intake; this, in turn, helps in the utilization of the poor-quality feed from the grass components. The introduced legume may also promote progression of the grass cover to a higher level, leading, for example, to the elimination of the undesirable *Imperata cylindrica* where it is a problem. Reporting at a seminar in Taiwan in 1962 on the possibility of using tropical legumes with appropriate inoculation, either in sown mixtures with grasses or for oversowing in stands of *Miscanthus* and *Saccharum spontaneum*, DAVIES (J.G.) stated (1970): 'This is not merely important to us in Australia and to you in Taiwan, but to the whole tropical and subtropical world. There are vast areas of potential pasture country, but first class pastures in these areas have not been developed because the problem of growing tropical and subtropical legumes has not been solved.'

Two species of *Stylosanthes* are those most used for the purpose of sod seeding, *S. humilis*, the annual, and *S. guyanensis* ssp. *guyanensis*, the perennial, which is more generally known under its synonym, *S. gracilis*. The literature on *S. gracilis* has been reviewed by TULEY (1968), who quotes reports of success in Zambia, Congo, Northern Nigeria, Cameroon, Madagascar and Malaysia, on introduction into natural grassland merely by broadcasting the seed on the surface of the soil; more reliable methods may involve some form of surface cultivation or sod-seeding.

STOBBS (1969b) reports on the effects on animal production in Uganda of oversowing natural pastures (mainly *Hyparrhenia rufa*, *Panicum maximum* and *Imperata cylindrica*) with *Stylosanthes gracilis* with and without a top-dressing of single superphosphate. The fertilizer increased herbage yields mainly by encouraging more vigorous growth of the legume; *Imperata* was suppressed in favour of the other grasses.

When East African Zebu steers grazing on a *Hyparrhenia rufa*/*Stylosanthes gracilis* pasture were moved to a new paddock, animals selected diet richer in crude protein than on a heavily grazed area (STOBBS, 1969e). The lower intake of protein is most critical at the end of the dry season when grass falls below 7 per cent crude protein, and when a marked decline in intake occurs. This may explain why *S. gracilis* is of most value in the dry season. The palatability of different parts of the plants requires study. Seeds of *S. gracilis* from faeces and fistular samples are viable and have higher percentage germination than seed collected in the field, probably due to scarification during digestion. Much leguminous seed is distributed in this way, although perhaps in an irregular pattern related to the grazing behaviour of the animals.

The following mixtures were sown in the Middleveld of Swaziland – predominantly summer rainfall, 760 to 1140 mm. (I'ONS, 1968):

Pennisetum clandestinum/Desmodium intortum Greenleaf and Silverleaf
Eragrostis curvula/Desmodium pilosiusculum/Stylosanthes guyanensis
Paspalum dilatatum
Setaria sphacelata c.v. Nandi/*Panicum maximum/Desmodium* Silverleaf.

Stylosanthes guyanensis was sod-seeded in Moist Tall Grassveld composed of *Cymbopogon validus*, *Rendlia altera*, *Paspalum commersonii*, *Eragrostis curvula* and *Loudetia simplex*, providing grazing for 7 to 8 months of the year at a summer carrying capacity of about one mature livestock unit per hectare (I'ONS, 1967, 1969). In the second year, weight gains in the treated areas were 64 per cent higher, or, in value, almost double the cost of annual applications of fertilizer. The legume may be damaged by indiscriminate burning and heavy grazing.

It is known that trials are in progress at the Indian Grassland and Fodder Research Institute, Jhansi, U.P., but results are not available (see SPAN vol. 15, p.l, 1972). CHATTERJEE, PREMCHAND & SINGH (1969) have reported on the growth of *Heteropogon contortus* swards sod-sown with legumes in Bihar. The Brunei Shell Petroleum Company finds at its Sinaut Agricultural Centre that, in addition to the introduced varieties of *Digitaria decumbens* and *Brachiaria brizantha*, the Nandi and Kazungula cultivars of *Setaria sphacelata* are promising (GROOME, 1972). Both these grasses have grown vigorously and are very palatable; but their erect, clumpy growth makes them unsuitable for grazing. It may be possible to sow *Pueraria phaseoloides* and *Centrosema pubescens* to provide ground cover and to climb up the erect stems of *Setaria*. The cattle would then take the legumes as they graze the *Setaria*; usually the local cows take the legume only after all other herbage has been consumed.

In coastal and subcoastal areas of Queensland, open sclerophyll forests, mainly *Eucalyptus* spp., contain a herbaceous understorey consisting mostly of perennial grasses and few perennial and annual legumes ('t MANNETJE, 1965). It is generally accepted that *Themeda australis* was originally dominant, but that continuous grazing by sheep and cattle combined with annual burning have reduced this palatable grass to that of a minor constituent. Its place has been taken by *Heteropogon contortus*.

Perhaps the first report on trials with pasture legumes in Queensland is that by SHAW (1961); a natural pasture of *Heteropogon contortus* was oversown with *Stylosanthes humilis*, and fertilized with a mixture of superphosphate, potassium chloride and molybdenum. The combination of increased carrying capacity and gain per head resulted in a five-fold increase in beef production per hectare. Subsequent studies (SHAW & 't MANNETJE, 1970) were designed to separate the effects of fertilizer, stocking rate and the legume, and to follow long-term changes in botanical composition ('t MANNETJE & SHAW, 1972). Overstocking of natural pastures in which *Heteropogon contortus* was dominant led to a reduction in its frequency, and increases in frequency, according to treatment, of: *Eremochloa bimaculata*, *Chrysopogon fallax*, *Tripogon loliiformis*, *Chloris divaricata* and *Digitaria longiflora*. These detrimental effects of overgrazing were prevented by fertilizer or by over-sowing with *Stylosanthes humilis*.

Given favourable weather (sowings followed by rain), the early establishment of *S. humilis* in uncultivated, uncleared, native annual or perennial pastures at Katherine, Northern Territory, is as good as on cleared cultivated land; equally good on clay loam and sand; soil surface treatments do not affect establishment. Mechanically dehulled seeds had germination two to three times higher than pods (MILLER, 1967; MILLER & PERRY, 1968). GILLARD (1970) reports on a grazing experiment on native pasture (*Heteropogon contortus*, *Bothriochloa intermedia*, *Chrysopogon fallax*) oversown with *Stylosanthes humilis* in the dry inland (rainfall 630 mm., evaporation 2,000 mm.) of northern Queensland, on land carrying one beast to 15 hectares in the unimproved state. These conditions are more arid than the coastal areas where *S. humilis* has been used for some years.

C.S.I.R.O. Division of Land Research at Katherine N.T. is studying the *Stylosanthes humilis*/annual grass system (Fig. 7/1), designed to gather data and develop concepts whereby germination, growth, competition and nutrient cycling can be modelled separately; later, accounting for the effect of grazing on these ecological components and their environment. A three-point component diagram (Fig. 7/2) illustrates the percentage distribution at the end of the wet season of areas dominated by legumes, grass or bare ground. This shows a serious move towards the bare ground pole prior to the sixth year of pasture. Pasture stability is defined as the aerial proportion of any given pasture which retains an unchanged type of ground cover at the end of one wet season when compared with that at the end of the previous season. Fig. 7/2 shows a fall in pasture stability from 69 per cent for the second and fourth year period to 39 per cent prior to the sixth year.

If suitable commercial dehullers were developed, *Stylosanthes humilis* would be distributed and sown as free, scarified seeds, rather than as pods as at present. A study has been made (L.A.T. BALLARD in

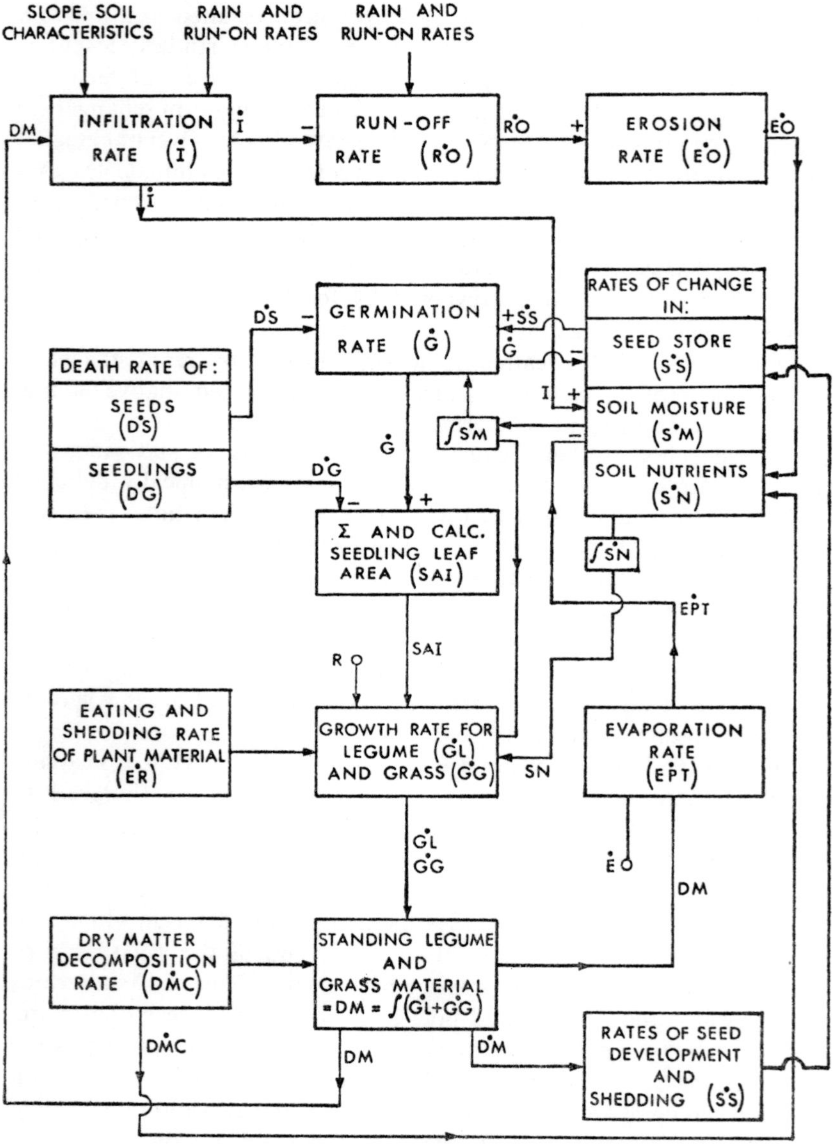

Fig. 7/1: Information flow diagram illustrating dominating components in *Stylosanthes humilis* – annual grass system. R, radiation; E, rate of transpiration. Arrows indicate transfer of variables.

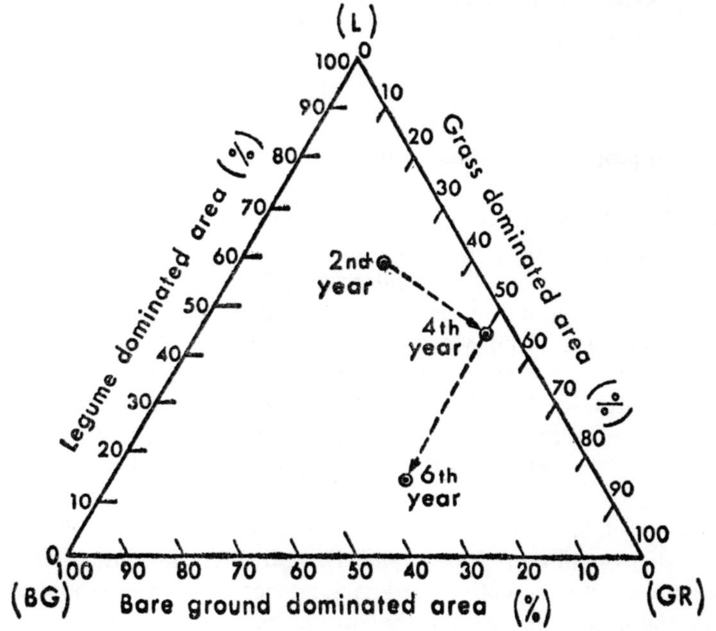

Fig. 7/2: Percentage distribution of ground cover at the end of the wet season of the second, fourth, and sixth year of a *Stylosanthes humilis* pasture.

C.S.I.R.O. Division of Plant Industry Annual Report, 1971) on extracted, scarified seeds, to obtain information about the embryo dormancy status, uncomplicated by restrictions imposed by pod or impermeable seed coats.

Considerable dormancy is still apparent four to six months after harvest, when seeds are incubated at constant temperatures. But seeds are particularly sensitive to single, brief exposures to small temperature shifts of 2.5 to 5 °C. The effects of the following variables on this response have been established: cultivar, post-harvest age of seed, length of imbibition period prior to the start of alternations, number of cycles required, magnitude of the temperature difference, and length of time at the lower temperature.

8 REPLACEMENT OF NATURAL COVERS

8.1 Small proportion of total area

It has been said that it is only under particularly favourable conditions of climate, soil and economics that one may consider replacing the natural plant cover of tropical grazing lands entirely, by clearance, ploughing and seeding or planting of adapted species of grasses and legumes (BOMMER, 1966; MOORE, 1970; WILSON, 1968). It may be possible in State-owned land or farms, or where ranching companies are operating (MOFFAT, 1966). In limited areas, ecoclimatic and economic conditions are favourable, e.g. Queensland, Jamaica (RICHARDS, 1970) and perhaps other places in the Caribbean and Central America, parts of Brazil (ROCHA, 1972), the highlands of Kenya (BOGDAN, 1959; MORRISON, 1966, 1969), other parts of east Africa (STOBBS, 1969d), Ghana (KANNEGIETER, 1966), and Hawaii (PLUCKNETT, 1970). Elsewhere, one cannot expect the ordinary grazier to be willing and able to provide the energy and capital that would be necessary, no matter what the urban-based departments and their well-meaning research agronomists and field advisers may recommend as desirable for the national good. It is difficult to see any long-term or permanent role for sown pastures in India (WHYTE, 1971), in Sri Lanka (ANDREW & JAYAWARDANA, 1970), or in Taiwan (DAVIES, 1970), in spite of the work of the UNDP/FAO project on marginal lands at Tainan, or in central Laos (THOMAS & HUMPHREYS, 1970). It will be interesting to watch the progress of the cattle ranches to be established in south Sulawesi, Sumba and Timor, and perhaps later also in Sumatra and Kalimantan, to be financed by World Bank credits amounting to U.S. $ 3.6 million, plus additional amounts for a feed mill and slaughterhouses. The feed resources of these ranches will presumably be based on the planting or seeding of grazing pastures, since any natural grass covers must at present be primarily *Imperata cylindrica* or other species of low economic potential.

Within the cultivated lands of the countries of the intertropical belt (again other than Australia) it is not possible to expect a great increase in the area of sown pastures sufficient to represent a considerable proportion of the total cultivated area. Rather one may expect to see increasing evidence of that polarization to which WOODFORD (1971) and WHYTE (1971) have referred. In the intertropical zone in particular, the emphasis on cultivated and cultivable land must continue to be on the production of food crops for direct consumption by human beings and on cash crops important to the national economy.

Yet it is to the small percentage of land that is technically and economically suited for sown pastures and fodder crops in the intertropical belt that most of the effort in research and development is concentrated, and to which most of the published literature relates. This work does not fall directly within the terms of reference of the present contribution. It is, however, important where the cultivated land can be regarded as an adjunct to the uncultivated range, since together they may form an integrated type of management within economic ecosystems.

ANDREAE (1966) estimates that 'die Grünlandflächen' of the world represent twice the area of the cultivated land, but make a relatively small contribution to world food production. Discussing the tropical and subtropical developing countries of southern Africa, it is argued that the transformation of free-range grazing land into productive animal husbandry is possible only where industrialization provides mass revenue to raise the agrarian price level, or when the infrastructure of land opens the country to trading in land. Thus the difference in price and cost between producers near to and far from markets diminishes.

8.2 Human sociological distinctions

Over much of the intertropical zone, animal husbandry and agriculture are the responsibility of different departments of government, and may be practised by ethnically different populations. Responsibility for research and development is also allocated to different bodies. In those African countries in which French overseas research specialists are operating, the Institut de Recherches agronomiques tropicales et des Cultures vivrières (IRAT) is not concerned with the natural grazing lands, which are for the herdsmen a gift from the gods on which they are most unwilling to sink any capital. These are the responsibility of I.E.M.V.T. (BILLE, LEBRUN & RIVIÈRE, 1968/9; PAGOT, 1971).

A report for 1970 of the work of IRAT in Cameroon, Dahomey, Madagascar, Réunion and Mauritius, and Senegal, is published in *Agronomie Tropicale* no. 1 for 1972. IRAT is concerned not with the stockbreeder, but with the cultivator who uses his animals for draught, for intensive cultivation or planned rotation with a rest period or fallow (BORGET, 1971). The first stage of improvement is the enriching of this fallow, if it is long-term, by sowing appropriate species, generally legumes; the second, the sowing of temporary pasture or ley, which may have a marked effect on the maintenance of soil fertility in the rotation (STOBBS, 1969a); the next stage would be permanent fodder cultivation, not part of the rotation, especially for summer fattening and under irrigation.

It is in similar sequences elsewhere in the tropics that polarization may be expected, leaving only the natural grazing lands and the most intensive forms of fodder cultivation and conservation (CATCHPOOLE & HENZELL, 1971).

8.3 **Changed conditions of soil and habitat**

The ecology of the pure sown or planted stands of grass or of the synthetic grass/legume mixtures is entirely different from that of the communities on the grazing lands. The soil structure has changed from a grassland or a grassland-cum-forest type soil to a cultivated or arable type soil. All competition of wild tree, shrub and herb associates has been removed. The plants which are used are 'improved' agrotypes of the wild species found in nature, and are thus better able to respond to the artificial conditions of nutrition (WILLIAMS & ANDREW, 1970; HENZELL, 1970) and soil water status which are provided by the cultivator (see also DAVIES & JONES (1971a and b) on 'Dauerweiden' and 'Ackerfutterpflanzen' in the tropics and subtropics; BOMMER, 1966).

8.4 **Tailoring species for cultivation**

The selection and breeding of species for sowing as short- or long-term pastures now demand a new approach, with special attention to the requirements of the animal that will utilize the artificial pasture, to the evaluation of new cultivars in genetical, physiological and biochemical terms, and to the performance of breeders' material in intense field competition and a reasonably well-defined environment (THOMAS, 1966; CORKILL, 1966; MORLEY, 1962; HUTTON, 1965, 1971a). Apomixis, cytoplasmic male sterility and/or vegetative propagation may make the large-scale use of high-yielding F_1 hybrids economically feasible (BURTON, 1970). For an environment such as the semi-arid areas of Kenya (630 to 890 mm. rainfall), plants must have a long life-span, resistance to seasonal drought, persistence, aggression in establishment, and absence of need for expensive protection against weed competition (SANDS, THOMAS, KNIGHT & PRATT, 1970).

In Chapters 9 and 10, we turn to the origin of the plant families, Gramineae and Leguminosae, and to the evolution of the many genera and species which contribute to the grazing and fodder resources of the intertropical zone. No distinction will be made between those plants which contribute in the wild state and those which have become 'crops' after collection, introduction and improvement for the special ecological conditions of cultivated land.

9 THE GRAMINEAE

9.1 Importance

There is no doubt that the Gramineae are the most important plant family in both the scientific and economic sense. They are pre-eminent on the score of absolute completeness of distribution (GOOD, 1964), and therefore excellent material for the study of taxonomic history and geography. Not only are members of this family found over the widest extremes of altitude and latitude, but their degree of distribution within this general range is particularly dense and continuous. Practically alone among flowering plants, the grasses now form a dominant element in the vegetation over great areas of the world, and almost everywhere else their proportion in the natural vegetation is high. HUBBARD (1968) estimates that there are about 620 genera and 10,000 species in the whole world.

The cereals provide the staple foodgrains for the nutrition of human beings everywhere. The wild and cultivated grasses provide most of the grazing and fodder resources upon which domestic and many wild animals depend for their sustenance and productivity. In these respects, they are followed by the Leguminosae. In developing a global strategy for human survival, it is desirable that as many scientists as possible concentrate and co-ordinate their efforts on these two plant families, at all levels from the most academic to the most practical.

9.2 Origin

9.2.1 CONTINENTAL DRIFT AND GEOLOGICAL HISTORY

General views on continental drift in the Pacific and eastern Asia in relation to the origin of the angiosperms have been given by MELVILLE (1966). The availability of additional evidence has necessitated some modification of these earlier conclusions (MELVILLE, 1972) (see also 9.5.4).

The limiting date for the time of origin of angiosperm families is the period at which Africa plus South America was separated from the rest of Gondwanaland. This, according to palaeomagnetic evidence, was at the beginning of the Triassic (MELVILLE, 1972).

'By this time, a number of the more important angiosperm families must already have been set on their evolutionary path; they were probably still gymnosperms at that date, but the general lines of their morphology and evolution had already been determined. I would place the Gramineae in this group. This means that one set of grasses would have been separated in Africa and South America, whereas the remainder would have been able to colonize or migrate into India and Australia, and by Jurassic times at least to have migrated to the land masses then existing in the central Pacific . . .. I am now satisfied that at least the northern half of Borneo belonged to the Pacific continent. The southern half probably belonged with some other parts of Indonesia which were associated with India before the drift movements took place.'

The family Gramineae has been said to be sufficiently large, diverse and old to reflect nearly the whole of angiospermous history (BEETLE, 1971). In the Mesozoic, the Tethys Sea divided the Old World climates, thus establishing the striking contrasts between the panicoid/chloroid and the festucoid types of grasses, the first being predominantly tropical, the second predominantly temperate.

'Some species, some genera and some tribes have achieved stability and some floras their characterization much earlier than others. As the classification of the grasses becomes more natural, distribution patterns will help to reveal the historical development of these groups.'

BROWN & SMITH (1972) relate the Kranz syndrome (see 12.3), utilization of available CO_2, and continental drift to the evolution of some families, subfamilies and genera within the Order Gramineae. Their conclusion is that the Eragrostideae and the Aristideae evolved from some late Jurassic to mid-Cretaceous grass types, probably tropical, that very early in their history acquired among their characters the Kranz syndrome incorporating the C^4 pathway. At the time of the breakup of Pangaea, these types were already in Australia, India, Africa and South America, where they later expanded greatly. BROWN & SMITH consider that the geographical distribution of the Eragrostideae is a classic example of the biogeographical proof of continental drift.

The following paraphrased and greatly abridged translation from Chapter V of the monograph of JACQUES-FÉLIX (1962) presents one interpretation of geobotanical history of the Gramineae of Africa, in relation to those of other continents. This should be considered together with the propositions of AUBREVILLE (1962) (3.1.1.) and PRAT and others (9.2.2 below).

In the Tertiary, when the equatorial zone extended over Europe, there were corresponding climatic zones in Africa. When these moved towards the south, most of the ancestral Gramineae had already been differentiated. The Bambuseae, Arundineae and Paniceae already existed in the various biotopes of the forest zone, while the Andropogoneae appeared on the forest borders between the equatorial and the torrid zones.

The differentiation of xerophytic types, the Eragrosteae, Chlorideae and Aristideae occurred in the tropical zone of Capricorn which then covered the north of the African continent; in the extra-tropical zone more to the south, the Danthoneae developed. In the course of climatic migration, plants of all these tribes were able to reach America, Africa and Asia. The spread of the ancestors of the Festuceae from their centre of origin (*Streblochaete* and *Pseudobromus*) occurred only when climatic migration created a cold zone over the higher continental areas. The Oryzeae migrated with the intertropical zone while the Phalarideae became adapted to the advancing microthermic conditions. Towards the end of the Tertiary, when the situation in Africa was as it is at present, the very ancient Paniceae (and the Chlorideae and Eragrosteae) developed original xerophytic types in their wide southerly spread over the whole extent of the African continent. JACQUES-FÉLIX (*op. cit.*) is most interested in the enclaves of holarctic species to be found in the mountains of Africa; the Cameroons has six genera of Festuceae, the Angolan plateau only one. Distribution of the Agrostideae was favoured by a greater facility for dissemination and adaptation to the mesothermic environment of the high plateaux.

Certain genera of the Paniceae are not pantropical but are evidence of an ancient climatic and geographical phase. *Cyrtococcum* and *Ottochloa,* which belong to the fairly closed environment of the deep forest, also occur, according to JACQUES-FÉLIX, in similar biotopes in monsoon Asia and Malaysia, their distribution having had the same Tertiary history as the forest element to which they belong. Others like *Holcholemma* and *Entolasia,* with only a few species, exist in Africa, India, Sri Lanka and Australia.

Different genera of the Andropogoneae are common to America and Africa, more frequently to Africa, Madagascar and Asia, or to the three continents and Australia. In the Afro-Asiatic genera, there is a considerable ecological disjunction for the species of the humid regions; these communities may be attributed to Tertiary migration of the palaeotropical element. Several of these elements have a restricted distribution in Africa, e.g. *Capillipedium parviflorum,* widespread in all the lands of the Indian Ocean, remaining isolated in East Africa. There are many intertropical elements which penetrate easily into the temperate region. Others like *Erianthus, Dichanthium* and *Bothriochloa,* spread as easily in the warm temperate regions as in the intertropical. Sometimes a single species, e.g. *Andropogon distachyus,* penetrates the intertropical region only on mountain summits.

Following the arguments put forward in 2.3 and 9.5.1, and the classification of south-east Asian Gramineae adopted in 9.5.2, one questions whether JACQUES-FÉLIX and other geobotanists give enough credit to the history and extent of grass-less types of forest vegetation and to anthropic factors in mapping the present regional or global distribution of genera and species. In addition, it will be helpful if detailed studies of the archaeology (antiquity) and present taxonomic geography were also made on more individual species, as has been done by BOURREIL & GILLET (1971) for *Aristida rhiniochloa.*

9.2.2 TAXONOMIC APPROACH

It is generally accepted that the monocotyledons arose from the dicotyledons, and sometimes that the Bambusoideae are the ancestral group of the Gramineae. However, many species of bamboo also show specialization, and a common ancestor may be a more acceptable proposition. The criteria for primitiveness depend upon individual phylogenetic theories. The Russian interpretation (Fig. 31 of TAKHTAJAN, 1969) is that the grasses (Poales) are 'the final link in a chain of anemophilous evolution, some basic stages of which may already be seen in the order Commelinales. By all accounts they arose directly from the Restionales, most likely from some extinct Flagellariaceae believed to have originated in highland areas with a low incidence of insect pollinators' (CVELEV, 1969). The Bambusoideae separated first, followed by the Oryzeae, Arundinelleae (or Arundineae), Eragrostideae, Panicoideae including Paniceae and Andropogoneae (see JAUHAR & JOSHI, 1966; 11.5), and finally Festucoideae. In the last group, further evolution is seen to have been associated with a reduction in the basic chromosome number, development of an annual habit, and simplification in structure. There is a general impression that the Andropogoneae are derived from Panicoid stock, and are hence less old (VELDKAMP, 1971).

PRAT (1960) has given a synthesis of his ideas with those of STEBBINS (1956, 1969), BEETLE (1957), TATEOKA (1957a, 1957b), and BROWN (1958). PRAT approves of STEBBINS' conclusion that no group within the Gramineae can be considered as 'original'. The Bambusoideae cannot be regarded as the ancestors of the family; although their flowers and fruit may have primitive characters, their anatomy and their epidermes are highly evolved. Their woody structure is certainly a secondary character, the common ancestor of the Glumiflores and the Liliiflores being probably herbaceous. A hypothetical mark X has therefore been included by PRAT (Figs. 9/1 and 9/2) to indicate a point in the Cretaceous when the different groups may have diverged, but unequally and not in the same manner. Differential rates of development of characters are found, the Festucoideae having evolved more rapidly in flower and epidermis characters, and Bambusoideae showing the converse situation.

Reference may also be made to the symposium on the natural classification of the Gramineae held at the Eleventh International Botanical

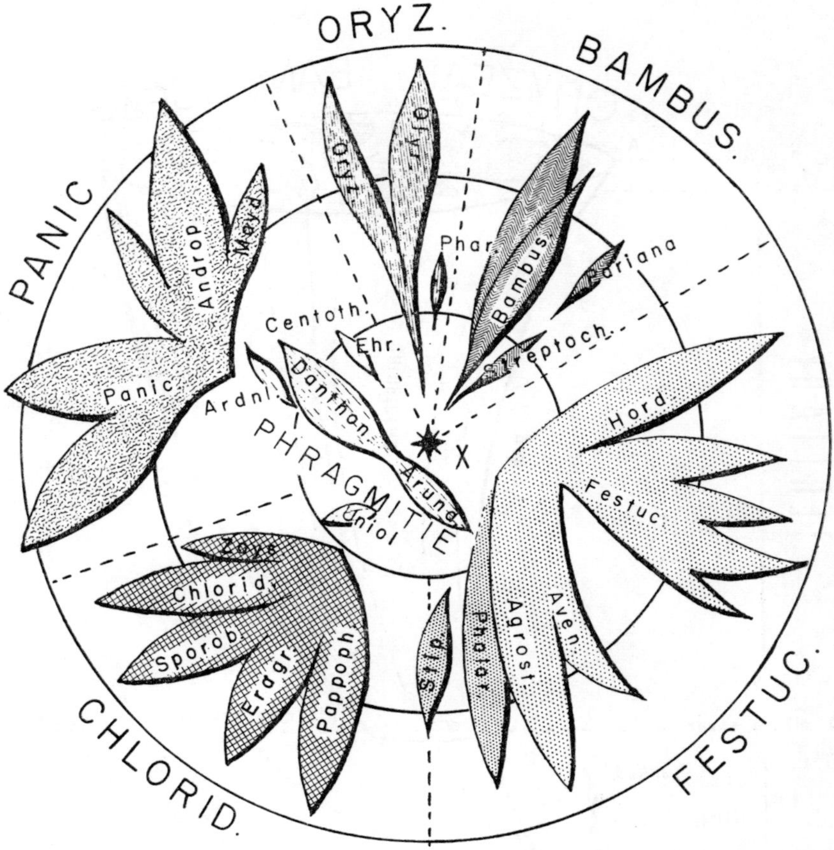

Fig. 9/1: Diagram showing the relations between the sub-families of the Gramineae, according to various specialists, summarised by PRAT (1960). The star in the centre indicates a hypothetical ancestor, now extinct.

Congress, Seattle, Washington, 24th August to 2nd September, 1969 (I.U.B.S., 1971):

J. R. REEDER, The grass embryo in systematics

I. ROTH, Histogenesis and morphological interpretation of the grass embryo

H. PRAT, Emploi des caractères épidermiques dans la classification des Graminées

T. TATEOKA, On the phylogenetic relationship between *Lepturus* and *Monerma*

F. W. GOULD, Polyploidy in grasses of southwestern United States

W. V. BROWN, Grass leaf anatomy: its use in systematics

R. C. LOMMASSON, Grass leaf venation

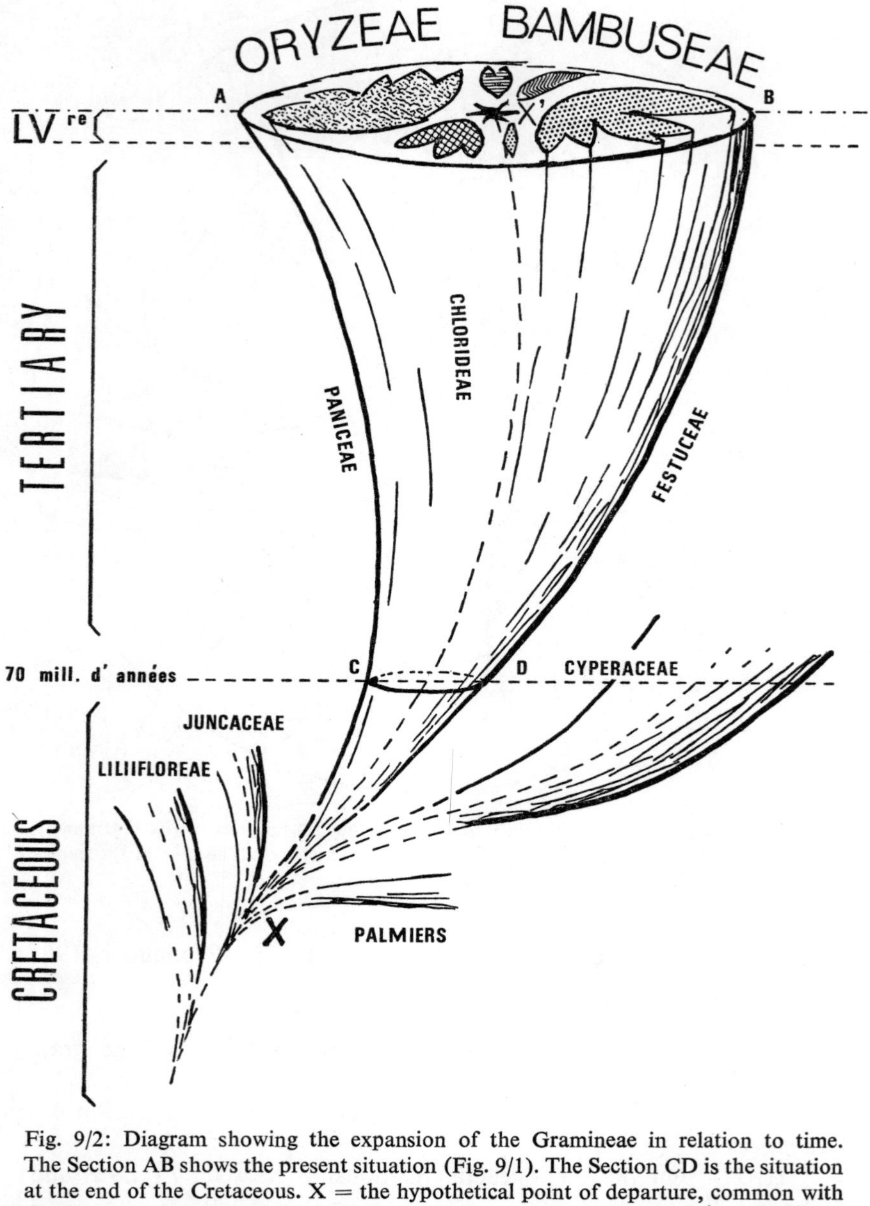

Fig. 9/2: Diagram showing the expansion of the Gramineae in relation to time. The Section AB shows the present situation (Fig. 9/1). The Section CD is the situation at the end of the Cretaceous. X = the hypothetical point of departure, common with the Cyperaceae, and near those of the Juncaceae and the Liliifloreae. X = projection of the present situation indicated in Fig. 9/1.

A. A. BEETLE, Distribution as a key to the age and origin of grasses
D. E. FAIRBROTHERS & M. A. JOHNSON, The precipitin reaction as an indicator of relationship in some grasses
K. L. MEHRA & E. ANDERSON, Character association analysis: its use in grass taxonomy
L. R. PARODI, La taxonomía de las Gramineae Argentinas a la luz de las investigaciónes más recentes
J. M. J. DE WET, Phylogenetic position of some endemic South African grass genera
G. L. STEBBINS & B. CRAMPTON, A suggested revision of the grass genera of temperate North America.

9.2.3 GEOGRAPHICAL APPROACH

It is necessary to re-examine the proposal of BEWS (1929) that the progenitor of the grasses arose in the humid tropical forests, and that migration to less uniform, more exacting and variable environments was accomplished by mutation and natural selection. Modifications of various kinds in the primitive progenitor would ensure greater protection for the seed and a complicated dispersal mechanism as the present-day end-products of trends of evolutionary change (BOR, 1960). On present evidence, one may propose that the place of origin of Asian mesophytic grasses was in a zone of vegetational transition (VALENTINE, 1970), in some type of forêt claire, away from the tropical rain forest zone proper; and that the primitive grasses migrated in two directions 1. towards the equator and into the few ecological niches there available to them, and to become insect pollinated (SODERSTROM & CALDERÓN, 1971) and distributed by adhesion of seeds to animals as the only means of migration on the windless floor of the forest (ARBER, 1935), and 2. out into the more xerophytic world (but see also 5.3).

9.3 Taxonomy

9.3.1 ROYAL BOTANIC GARDENS, KEW

Since the field grassland worker is frequently somewhat mystified by the techniques and objectives of the grass taxonomists, it is desirable to give the latters' point of view at some length, by a quotation from CLAYTON's presentation (1966) of the views of the Royal Botanic Gardens, Kew, to the Ninth International Grassland Congress in Brazil:

'In the tropics, the correct names for quite common grasses are often still in doubt. The grassland worker soon finds himself in the shifting sands of an unsettled taxonomy, and some acquaintance with the elements of the subject is essential if he is not to become entangled with ambiguous names .... Currently research on the taxonomy of tropical grasses is very active, and the following is a brief account of the main reasons for those name changes which the field worker finds so tiresome.

'The naming of plants is governed by the International Code of Botanical Nomenclature, which sets out the conditions to be fulfilled when a new name is given, and which regulates the choice to be made when the same plant has received several different names. It is not exceptionally difficult to understand and has certainly brought order out of anarchy ....

'Perhaps the most misunderstood section of the Code is that concerning the type concept. At one time the meaning of any name could be determined only by the description supplied with it, however inadequate that might be. Nowadays the Code directs that a name shall be permanently associated with the herbarium specimen from which the original description was drawn – the type specimen. By reference to the type specimen the description can be augmented whenever doubts arise. For example Linnaeus did not foresee that several species closely allied to *Sporobolus indicus* (Linn.) R. Br. would be discovered, and his description does not enable us to distinguish between them. Examination of his own specimen however soon reveals which of these species should continue to bear the epithet that he has bestowed. It should be emphasized that the type specimen is in no sense typical of the species, nor is it the pivot around which the species is constructed. It is simply the specimen to which a name is attached for nomenclatural purposes, and it is important to realize that taxonomy and nomenclature should be treated as separate subjects. The major part of a taxonomist's task is to sort his material into populations representing species. It is only when this is complete that he seeks out the type specimens for examination. Should one of them fall within the range of variation of a population it at once gives it a name; sometimes a population embraces several types, and a name is chosen from among them according to the rules of nomenclature; rarely there are none and a new name must be given ....

'It should not be forgotten that the object of formal taxonomy is to provide a general purpose classification based on overall resemblances between as many characters as possible. Many alternative classifications are possible, such as those based on chromosome number or breeding system, and anyone is free to devise a more suitable scheme for his particular purpose. Unfortunately, the originators of special purpose classifications often feel obliged to invest them in formal latin nomenclature, thereby making unalterable additions to synonymy. An informal system expressed in common language is often preferable, for, besides being more flexible, it leaves the general purpose classification unscathed.

The increasing use of the Horticultural Code for classifying agricultural plants is an example to be emulated' (See also 11.2).

9.3.2 Numerical Taxonomy

Assistance is being given by the CSIRO Division of Tropical Pastures (W. T. Williams, in Annual Reports of Division for 1969/70 and 1970/71) to the Division of Computing Research (G. N. Lance) and the University of Queensland (H. T. Clifford) in reviewing the existing battery of taxonometric programmes for the Control Data 3600 computer. A new algorithm has been devised for the comparison of two classifications of the same set of elements. A major new project is the development of a polythetic divisive strategy, with reallocation and allocation of facilities. The basic procedure is a prior monothetic division followed by a reallocation on a likelihood model.

'New models are being explored for the classification of highly heterogeneous data, in which a large proportion of the possible inter-individual comparisons represent completely dissociated pairs. The most promising model uses the Bray-Curtis measure with the constraint that no group may contain a dissociated pair; the hierarchy is completed by progressive relaxation of the constraint. The model has been used successfully on intractable data from the Division of Plant Industry and on marine survey data from the University of Queensland; consideration is being given to developing it into a major production program.

'There has been the usual wide variety of semi-routine data-processing for a number of workers inside and outside the Division. Three aspects of the local work have been particularly prominent:

1. Analysis of vegetation surveys of grazing experiments, aimed at elucidating inherent site-patterns which may affect the experimental results. The program CENTPERC at Canberra, now widely used by investigators elsewhere, was originally designed for use in this Division.
2. Analysis of plant introduction material; work has been carried out on material of *Stylosanthes*, *Urochloa* and *Phaseolus*. The most profitable form of analysis has commonly been classification using the Burr Euclidean model, followed by a principal co-ordinate analysis.
3. Discontinuous methods of analysis (manifold contingency analysis and multiple predictive analysis) have been introduced into the analysis of fertility data, and have proved effective and simple.'

In the Botany Department, University of Queensland, five computer programmes have been used to classify 92 genera on the basis of 50 characters. Leaf and fruit characters were the most useful, spikelet characters of slight assistance in definition of groups (Clifford, Williams & Lance, 1969). Established correlations between attributes have arisen in part as a result of the phylogenetic origins of the genera, and in part as a response to natural selection determining distribution of genera adapted to present-day habitats (CLIFFORD, 1969).

9.3.3 CYTOLOGY IN GRASS SYSTEMATICS

The basic chromosome number is frequently 7 (festucoid grass; temperate regions and tropical mountains), 9 or 10 (panicoid-chloridoid grasses; warmer regions), sometimes 11, 12(6) or 13; polyploidy is not rare; chromosomes can be of different size, large in festucoid grass, small in panicoid-chloridoid; as with anatomical characters, intermediates occur (TATEOKA, 1960; see also SINGH, 1966). BROWN & SMITH (1972) assume that the lowest basic number was originally 6, in ancient grasses that are probably all now extinct; many of the less advanced tribes have the secondary basic number of 12, produced through Cretaceous or Jurassic lines with n = 13, 11 and 10; their modern descendants are believed to be the small temperate tribes with the basic number 10, 11, 12, 13.

9.3.4 FIELD COLLECTION FOR TAXONOMIC PURPOSES

The technique of plant exploration for economic purposes is discussed briefly in 11.6.3. No account has been found on the field technique to be adopted in the collection of material for use in taxonomic studies. How to sample an area in order to tap the full range of variability for that particular site? How and where to select a type specimen – in the field or in the herbarium? Particularly in the arid and semi-arid regions, species are changing under the impact of ever-increasing aridity. Material collected now may not fit the description of the species written 25 or 50 years ago. Anything approaching the herbarium type specimen can no longer be found in the field. Should new field surveys be made at intervals of 25 or 50 years, and new specimens recognized as types for the modern species or its successor, as they now exist in the field?

9.4 Speciation

9.4.1 MEETING IN NEW HABITATS

Taxonomists continue to report the discovery of new species. These may be either species in inaccessible places or which were missed in earlier collections. They may also be the result of renewed speciation. Following the emphasis placed in these chapters on the migration of species in relatively recent centuries and even decades, it follows that recognized species far from their habitat of origin now find themselves in rather dissimilar environments. This would seem to create conditions for intercrossing and renewed speciation in species which had already become somewhat tied to their original environment and somewhat fixed in their characters and degrees of variability. One must also consider whether the findings of the marine biologists regarding intertidal and shallow-water animals also apply to the Gramineae, namely that the tropics are regions of greater stress than the temperate zones, and that this may be one of the factors leading to high speciation in the tropics.

WHITTAKER (1972) has stated: 'Given a resource gradient (e.g. light intensity, prey size), in a community, species evolve to use different parts of this gradient; competition between them is thereby reduced. Given a habitat gradient (e.g. elevation or soil moisture conditions) species evolve to occupy different positions along this gradient. The various habitat gradients of a landscape may also be conceived as a multidimensional hyperspace, and species evolve toward occupation of different positions in this hyperspace. Along a particular habitat gradient species populations have scattered centers and usually overlap broadly, forming a community continuum or coenocline. Through evolutionary time additional species can fit themselves in along the coenocline. As they do so the extent of change in community composition along the gradient increases.'

ROSS (1972) equates the species diversity of a community with its ecological stability integrated with its geologic instability.

'Species diversity, here defined simply as the total number of species in a community, represents the historic processes by which additional species enter the community and species disappear from it. The additions arise chiefly through speciation by geographic isolation followed by a congregation of the new species formed. The mechanism is by reversible geologic events such as glacial periods, orogenies, and dispersal corridors, the first two affecting organisms as reversible climatic changes. These events cause speciation on a grand scale, acting simultaneously on species in all seral stages of the biome. On this basis, the input side of the species diversity of a community is a combined function of 1. the geologic longevity of its ecological conditions, and 2. the number of times the community area undergoes geographic disjunctions and reconnections.'

9.4.2 ECOLOGICAL REGRESSION INCREASES OPPORTUNITY

Most of the grass covers of the intertropical world are seral stages below the potential grass sub-climax for the particular environment.

Thus, species which were formerly present in very small percentages have now become dominant or sub-dominant. The mathematical opportunity for genetic proximity, intercrossing and change has been greatly increased. One would therefore expect to find much greater incidences of polymorphism resulting from intercrossing and leading up to the species level among the genera which consistently appear in low positions in the succession plants in Tables 5/2, 3 and 4.

9.4.3 Origin of Annuals

The taxonomic and historical origin of annual forms of the Gramineae is of fundamental importance in any study of the origin of cereals. Most botanists agree that annual forms of plant species arose from perennials. Annuality is a mechanism for escaping seasonally unfavourable environments. The change of a species population from the perennial to the annual state happens as an expression of the physiological stress produced by such environments (12.8) – great or increased intensity of desiccation and heat over long periods such as occurred at several periods and possibly in different places in geobotanical history, or adverse seasonal temperatures associated with higher latitudes and altitudes.

Perennials tend more towards vegetative growth, since reproduction by seed is not so essential to their continued existence. Annuals must produce large amounts of seed to allow for wastage, and the seeds are larger to help the next generation to become established in an inhospitable environment. The evolution of annual forms of specific or lower status from a perennial favours dispersal of that genus into new habitats in which the perennial forms could not grow.

The annual forms of the Gramineae are characteristic particularly of arid and semi-arid habitats, and are increasing in frequency with the progressive extension of man-induced desiccation. *Heteropogon contortus* is predominantly perennial in India, but an annual form is found in arid ecological niches on the Maharashtran Deccan. The occurrence of an annual form of this genus in the absence, until recently, of the perennials in Java (Monod de Froideville, 1968) is to be explained in terms of adventitious introduction. The more equable environments of the humid tropics are not conducive to the evolution of annuals.

Although annual species of Gramineae may occur in small percentages in arid and semi-arid communities, their presence as dominants in such habitats must be regarded with suspicion. They are then probably secondary and rather low in ecological regression. The climax grass communities around deserts or in arid and semi-arid zones generally are perennial; these growth forms may be progressively eliminated due to overgrazing, spasmodic cultivation and other causes, and replaced by less demanding secondary annuals of the same, but more commonly other genera.

9.4.4 Nature of Primitiveness

It has been said (Veldkamp, 1971) that a consensus on the morphological characters of a primitive species of the Gramineae is not likely to be achieved until the history of the angiospermous flower itself has been defined satisfactorily. Below (9.5.3), the historic significance of such terms as native, indigenous or endemic is discussed. Elsewhere, several different proposals are made in the historical context:

a. that the Gramineae of the intertropical belt contain most of the ancestral forms from which temperate tribes and genera have evolved (Chapman, 1970);
b. that the progenitor of the grasses arose in the humid tropical forests (Bews, 1929);
c. that the genera reported for these forests or forest fringes (2.4 and 9.5.2) may be expected to show primitive characters if, as Jacques-Félix (1962) has stated, they have always been associated with the forest type in which they are now found;
d. that many of the Asian mesophytic grasses had their place of origin in a more open type of forest, from which they would have migrated in two directions (9.2.3); thus the few grass species found at present within or on the fringes of tropical rain forest may be truly indigenous there, or may have migrated there from a subequatorial monsoonal habitat;
e. that some of the species found in high-altitude enclaves in the intertropical zone (9.2.1 and 9.5.4) might be expected to be primitive relics from an earlier climatic era; and
f. that in Africa (5.3) it is suggested that the primitive species might be found or might originally have existed in the contrasting mesophytic (? Paniceae) and xerophytic (? Andropogoneae, Chlorideae, Eragrosteae, Aristideae) palaeobotanic zones, and that they migrated from these habitats to their physiological limits of tolerance of aridity and humidity respectively, following palaeoclimatic movements of major forest types (3.1.1) or the much more recent reduction by man of grass-less forêts denses to forêts claires with sufficient light for a gramineous understorey, or to savanna.

9.5 **Taxonomic geography**

9.5.1 Floras and Statements of Distribution

The geographical approach to the taxonomy of the Gramineae calls for an analysis of the habitats in which the species are collected. Published floras are deceptive in this respect, since they record everything that is found, no matter where or in what quantity. They do not discuss why a species is found in a particular habitat, plant community, fallow field, roadside or graveyard. The cumulative effect of these activities is the presentation of a global or continental or ecoclimatic distribution of that species which is suspect and calls for further analysis.

By adopting the ecological/historical technique (2.1) it is possible to define the past, recent and current vegetation of an area being studied. If it can be proved that, in the recent past, it was a type of vegetation in which the species collected could not possibly have occurred, even in favourable ecological niches, then that site or those sites should be regarded as adventitious or secondary. By relegating reports of species distribution to this category, one may gradually eliminate much of the claimed distribution, and then narrow down the true, relevant distribution to the centre or centres of origin of that genus and species (11.1). The application of this approach to certain floras of Asia may be taken as an example of worldwide significance.

9.5.2 Grasses in Floras of South-East Asia

By applying the above method of analysis and deduction to Gilliland's grass flora of Malaya (1971), one obtains the following breakdown into indigenous, doubtfully indigenous and non-indigenous species:

INDIGENOUS

a. some bamboos
b. forest grasses adapted to shade of forest or relative shade of forest fringes (species of *Centotheca*, *Lophatherum* and *Leptaspis*; see 2.5 for global distribution of species of these genera; also possibly *Cyrtococcum* and *Ottochloa*, according to the remarks – 9.2.1 – of Jacques-Félix, 1962, Ch. 5); some species adapted to dispersal in the forest by having inflorescences or spikelets that become readily attached to the coats of passing wild animals;
c. moisture-loving and riverside species, of the genera *Oryza* and *Saccharum*, and possibly *Phragmites karka*
d. grasses of coastal and other sands, muds and estuaries (in part), some appearing to be indigenous (*Spinifex littoreus*, *Thuarea involuta*, *Oplismenus burmannii*, *Sporobolus virginicus* and possibly *Panicum repens*)
e. grasses of limestone hills (Henderson, 1939) in part, and
g. grasses of high altitudes above the forest

POSSIBLY INDIGENOUS: grasses for which habitat data are inadequate for historical conclusions; species of *Coelorachis*, *Ichnanthus*, *Isachne*, *Microstegium*, *Pogonatherum*, *Setaria*, *Sorghum*, *Themeda*, *Thysanolaena*.

NON-INDIGENOUS

a. grasses of coastal and other sands, muds and estuaries (in part)
b. grasses of limestone hills (in part)
c. grasses of north and north-west Malaya, extending into Thailand, grasses from western Monsoon Asia, of the genera *Apluda*, *Arundinella*, *Bothriochloa*. *Chrysopogon*, *Dichanthium*, *Dimeria*, *Eragrostis*, *Eremochloa*, *Eulalia* and *Neyraudia* (but, surprisingly, no *Sehima*), many or all of which were introduced by accident or intent, while the ancient trade routes were in operation across the Isthmus of Kra and thereabouts
d. ruderals of disturbed places, paths and roadsides, generally in good light; weeds of upland cultivation, shifting or settled, gardens, waste land, mostly recent arrivals during the past 100 to 200 years, as seed contaminants or in other ways: occurrence of *Phragmites karka* as a ruderal, not only on abandoned tin mines, but also along watercourses, suggests that it is not indigenous, but may have been introduced from the Indian subcontinent (also reported from streams in the Middle Andaman Islands), when the rivers of Malaya were used as inland trade routes
e. grasses of aboriginal village sites, their waste land, clearings in forest; much older than 200 years; a few escapes of primitive crops back to the wild, such as *Coix lacryma-jobi* and *Setaria italica*
f. grasses of padis, ditches, canals, grasses of wetlands and swamps; probably few if any are truly indigenous, and
g. introductions for economic or ornamental purposes.

Applying the same criteria to the Gramineae in the flora of Java (MONOD DE FROIDEVILLE, 1968) one finds that some 25 species may be indigenous, 68 species doubtful, and 182 non-indigenous foreigners which have arrived since the tropical rain forest was progressively cleared. The rate of colonization may have been rapid; the South American species, *Pseudechinolaena polystachya*, found for the first time near Bogor a century ago, has now spread over the whole of Java, at elevations from 200 to 1,650 m.

From an analysis of the twelve species of *Panicum* recorded for Malaya, it is doubtful whether West Malaysia and Singapore should be included in statements of their global distribution, except as constituents of a 'recent' adventitious flora. LAZARIDES (1971) has analyzed 11 species of *Panicum* found in the Malay Peninsula, Indonesia, Thailand, Indochina, Philippines and Papua-New Guinea, and concludes that 'on a critical evaluation of available data, I must admit that I cannot present any evidence to suggest that the Malaysian species are of ancient origin. It appears that a relatively recent origin might also apply to many if not all south-east Asian members of the genus.' A similar analysis is necessary for the sixteen species of *Panicum* recorded for China (KENG YI-LI, 1965), and for the twelve species recorded for Taiwan (HSU, 1971).

It may also be claimed that the eleven species of *Digitaria* in Malaya are not indigenous; VELDKAMP (1971) recognizes twelve species, and states: 'These, and many other grasses, are no doubt native and original in the region. No doubt human activity has given them a better chance for expansion, but I am sure that the abundance of contemporary individuals is of secondary importance in these studies.'

9.5.3 ENDEMISM

There is a need to consider a more precise meaning in historical and geographical contexts of the terms: native, indigenous and endemic. How long does a grass species have to exist in an area before it qualifies for the category of indigenous? An endemic species must presumably be one of three things in terms of geobotanical history: a. an ancient species which has been in that habitat for millenia, and exhibits characters of taxonomic primitiveness, b. a less ancient species which has evolved from a more ancient ancestor which may or may not now be extinct, or c. a species which has come from elsewhere, presumably before the advent of man.

According to HENDERSON's Dictionary of Scientific Terms: Endemic (Greek endemos = native): restricted to a certain region or part of a region

Indigenous (Latin indigena = native): belonging to the locality; not imported, native.

According to Shorter Oxford Dictionary: Endemic: peculiar to a people or a district; esp. of plants and animals; having their habitat in a specified district; opposite to exotic.

Indigenous: born or produced naturally in a land or region; native to (the soil, region, etc.).

WILLIS (1922) states that the term endemic has long been used to connote a species, genus or other group confined to a small area, such especially as a single island, a group of islands, a mountain chain or a comparatively small country; no real difference can be pointed out between endemic and non-endemic species (or genera); none of the features of distribution of endemics can be explained on the supposition that endemics are local adaptations or are relics; in general, they are young beginners. Neither the supposition that endemic and small genera are relics, nor that they are special adaptations, will avail to explain the phenomena presented by the great mass. Endemic genera further prove to belong more to the larger families, just as endemic species belong to the larger genera.

MERRILL (1906) recognized 46 species and varieties of Gramineae as endemic to the Philippines. The Flora Malesiana of the Rijksherbarium, Leiden, recognizes seven endemics in its region of study (VELDKAMP, 1971):

Monostachya oreoboloides at 2,500 to 4,700 m. in north Sumatra, north Borneo, southwest Sulawesi, Luzon and New Guinea; possibly a dwarf species of *Danthonia*

Ectrosiopsis eragrostoides

Ramosia sp. = *Centotheca*

Asthenochloa-monotypic

Polytrias-monotypic, doubtfully distinct genus

Sclerachne-monotypic; on heavy, periodically inundated soil, in light shade, open teak forests, padi fallows, lawns, road margins, up to 300 m. in Java, Kangean Islands, Bali, Sumba, Timor.

Buergersiochloa – two species, extremely rare, only few specimens collected in forests of New Guinea – habit bambusoid; one species has three lodicules; regarded by Rijksherbarium as apparently distinct, and of great phylogenetic interest.

RAO (1972) has included thirty-six species in twenty-one genera (seventeen monotypic) among his angiosperms endemic to the Indian floristic region and its neighbouring areas.

JACQUES-FÉLIX (1962, Ch. 5) notes the degree of endemism in the tribes of the Gramineae occurring in Africa. His map of the 18 endemic genera of the Paniceae shows the absolute poverty of the whole Sudano-Deccanian region. The Guineo-Congolese region is better represented with the typical forest genera such as *Microcalamus* (see BUTZIN 1970a and 1970b; 11.1) and *Commelinidium*, or the orophilous *Cloachne*, or the hydrophilous *Louisiella* and *Thyridachne*. The Zambezo-Abyssinian region is richest, with pronounced 'domanial' endemism, since the genera of the Abyssinian domain are not those of the eastern domain, nor those of the Angolan or Kalahari domains.

The endemic genera of the Andropogoneae are essentially distributed in the Sudano-Deccanian and Zambezo-Abyssinian regions; they encircle the Guineo-Congolese region which they enter by certain special biotopes. These are not in fact forest species, but plants like *Jardinea* which like marshy formations of great rivers, or *Urelytrum*, which spreads in the edaphic and anthropic savannas (JACQUES-FÉLIX, 1968; 5.9.3). Unlike the Paniceae, some genera are original in the Sudanian region: *Anadelphia*, *Monium*, *Dybowskia*, and many others which are common to the great amphi-equatorial region.

9.5.4 High Mountains in Tropics

The geoecology of the high-mountain regions of Eurasia, including the Himalayan region situated along the northern border of Monsoon Asia, has been reviewed by a group of specialists (Troll, 1972); included are: the worldwide differentiation of high-mountain ecosystems (C. Troll), the upper limit of aridity and the arid core of High Asia (C. Troll), the high valleys of the Mount Everest region in Khumbu Himalaya (W. Haffner), the three-dimensional zonation of the Himalayan system (C. Troll), and the eastern marches of High Asia and the river gorge country of the eastern Himalaya and further east-south-east Tibet, Yunnan and Szechuan (U. Schweinfurth).

The distribution of Gramineae in the high-altitude enclaves in the midst of the intertropical environment itself calls for synthetic study and analysis (Jacques-Félix, 1962, Ch. 5; Kenworthy, 1966; Troll, 1968a; Weischet, 1969; Hedberg, 1971). The species usually belong to genera which are now regarded as of temperate adaptation. There may be four explanations for their occurrence;

- (i) relict species from a Gondwanaland grass flora in a temperate climate,
- (ii) relict species following major tectonic disturbances,
- (iii) temperate relatives of nearby tropical representatives of the same genus, and
- (iv) species brought from temperate latitudes by migrating birds.

See 10.2 for parallel consideration of the Leguminosae.

The 'islands' of grass species of temperate physiological relations on top of many mountains of mainland and especially insular south-east Asia still await an acceptable explanation (Whyte, 1973), probably to be related to studies on the evolution of the Afroalpine flora (Hedberg, 1971), and of the Antilles (Howard, 1971), among others. Löve (1963) has reviewed the roles of sea and land birds in the dispersal of plants of the north Atlantic biota. Carlquist (1967) distinguished seven presumed modes of arrival of plants in the Pacific islands; those relating to dispersal by birds are most relevant to a biological explanation of the grass flora of south-east Asian mountains – seeds attached to feathers, eaten and carried internally, embedded in mud on feet. Bird migration from the northern hemisphere to the southern and back in the western Pacific will require investigation, e.g. terns, godwits and the long-tailed cuckoo. Do they fly in a high arc over the equator to avoid the tropical atmosphere; do they rest on the high mountains of Taiwan, the Philippines, Malaysia, Indonesia, Papua and New Guinea; what was their diet at points of departure and along the route of migration; at what season of the year do they depart and are grass seeds then available in adequate quantity; what is the content of their digestive tracts in flight;

how long is ingested food retained; how long do undigested seeds remain viable (PROCTOR, 1968)? The study of altitudinal distribution of land birds in New Guinea (KIKKAWA & WILLIAMS, 1971) is relevant.

The conclusion of MELVILLE (1972) has a direct bearing on the migration of many plant species: 'in particular the high mountain plants of Kinabalu (Sabah) have their relationships in the Pacific and New Zealand; although Kinabalu is comparatively recent in origin, there is evidence that another high mountain existed in what is now Borneo at an earlier stage, and has since been eroded down'.

Further evidence on the gramineous flora of the mountain enclaves in the intertropical belt (and therefore composed of short-day ecotypes) is readily available for Africa and Latin America. For example, on the native pastures in the Puna zone of the Central Sierra of Peru (3,700 to 4,300 m. elevation) FAO field officer I. BLAIR recorded the following: *Calamagrostis* spp. including *C. vicunarum* and *C. heterophylla*, *Bromus pitensis*, *Muhlenbergia* spp., *Poa* spp., *Stipa* spp. including *S. mucronata S. depauperata* and *S. ichu*, *Festuca* spp., including *F. compressifolia*, and *Agropyron attenuatum*.

Exotic species have been established on this land by sod-seeding or cultivation combined with application of fertilizers – 80 kg. N, 100 kg. P_2O_5 and 80 kg. K_2O per hectare. *Lolium perenne* and *Trifolium repens* are the most promising species: *Festuca pratensis* is also possible, but *Dactylis glomerata* and *Festuca rubra* are inferior to *Lolium perenne*. (E. PAHL & C. BOJÓRQUEZ, UNDP/FAO).

9.6 **Contributions to taxonomic geography**

A few studies selected at random and not in any way representative of the literature will indicate the types of information of particular value to those concerned with the constituent gramineous genera and species of tropical grazing lands:

Aristida: ecology, germination and cultivation of some African species (BOURREIL, 1967; BOURREIL & GILLET, 1971)

Axonopus: accidental arrival in West Africa from South America and rapid evolution following natural hybridization, spread to Australia and Malesia (GLEDHILL, 1966)

Brachiaria: seed viability (GROF, 1968, 1969b); cytology and taxonomy (SOTOMAYOR-RIOS, SCHANK & WOODBURY, 1970); *B. brizantha* pastures in Madagascar (GRANIER & LAHORE, 1966)

Cenchrus: *C. ciliaris* at Central Arid Zone Research Institute, Jodhpur, Rajasthan (WHYTE, 1964, 1968b), in Queensland (BURT, 1968) and Central Australia (WINKWORTH, 1963)

Chloris: review article on *Ch. gayana* (BOGDAN, 1969); frost susceptibility in *Ch. gayana* (JONES, R. M., 1969)

Table 9/1

Temperate Grasses at High Altitudes in Sabah (East Malaysia) and Indonesia

Kinabalu, Sabah 1.	Elevation	Kinabalu, Sabah, Sandakan Herbarium 2.	Elevation	Malesia 3.	Elevation
Species	metres	Species	metres	Genera	metres
Deschampsia flexuosa					
var. *ligulata* Stapf	3,140	*Agrostis canina*	2,900	*Agrostis*	1,300–1,400
Bromus kinabaluensis	1,800	*A. infirma*	2,950	*Anthoxanthum*	2,000–3,500
Jensen		*A. reinwardtii*		*Aulacolepis*	2,400–3,000
Anthoxanthum horsfieldii		var. *borneensis*	4,000		
Maxim.	to 4,000	(Stapf) Ohwi		*Brachypodium*	1,700–3,600
Poa epileuca	3,000 to	*Bromus* cf. *Anthoxanthum*		*Bromus*	1,700–3,560
	4,000		2,900	*Danthonia*	2,400–3,800
Isachne spp.	1,800 to	*Bromus* sp.	3,000	*Deschampsia*	2,000–4,000
	2,500	*Isachne elementis* Merr.	2,400	*Deyeuxia*	1,500–4,00
		I. cf. *kinabaluensis*	2,600	*Dichelachne*	2,400–3,680
		Monostachya centro-		*Festuca*	1,600–4,050
		lepidoides	3,600	*Helictotrichon*	1,700–3,500
		Trisetum paniceum (L.)		*Hierochloe*	2,100–3,700
		Richter	3,000	*Microlaena*	1,600–3,700
				Monostachya	2,700–4,100
				Muhlenbergia	1,300–2,400
				Poa	2,000–5,000
				n.b. *Poa annua*	
				(introduced)	900–1,300
				Streblochaete	1,800–2,800
				Trisetum	2,000–4,000

1. MEIJER (1965), from literature, especially STAPF (1894), 2. COCKBURN (1969) (names not verified) 3. VAN STEENIS & MONOD DE FROIDEVILLE (1968) n WHYTE (1968a).

Cynodon: origin and taxonomic distribution of species (HARLAN, DE WET & RAWAL, 1970a and 1970b); 5.6; genus in Nigeria (CHHEDDA, 1971), and in India (JAIN, 1967)

Dichanthium annulatum: references in WHYTE (1964); interspecific hybrids (REDDY & D'CRUZ, 1968: D'CRUZ & REDDY, 1969); see 11.4

Digitaria: chromosome numbers for 107 accessions at Cunningham Laboratory, Brisbane (GOULD, 1971); review of literature on Pangola grass (NESTEL & CREEK, 1962); Pangola grass in Florida, the Caribbean Islands and Central America (HODGES *et al.*, 1967; WILSON, 1959; ADENIYI & WILSON, 1960; SHAMBULINGAPPA, 1970; DEGRAS, 1968; OSBOURN, 1969; WHYTE, 1959); *Digitaria umfolozi* in the Sudanian zone of West Africa (BOUDET, RIVIÈRE, CLÉMENSAT, PAGOT & LAHORE, 1961).

Entolasia: the performance of the high-protein species, *E. imbricata*, in Kenya (MWAKHA, 1970; DOUGALL & BOGDAN, 1958; BOGDAN, 1963)

Heteropogon: *H. contortus* pastures in Queensland (TOTHILL, 1966, ISBELL, 1969); factors affecting relative dominance of *Heteropogon contortus* or *Themeda australis* – efficient seed burial mechanism gives establishment advantage (TOTHILL, 1968); see also 11.5; fertilizer response of indigenous pasture in relation to rainfall ('t MANNETJE & SHAW, 1972)

Panicum: growth and development of *P. coloratum* in Queensland (LLOYD, 1970, 1971); frost susceptibility of *P. coloratum* (JONES, 1969); studies at ORSTOM (Côte d'Ivoire and Paris) on taxonomy, cytogenetics and reproduction of *P. maximum* (11.5).

Paspalum: comparison of seventeen introductions and naturalized *P. dilatatum* in Queensland (SHAW, ELICH, HAYDOCK & WAITE, 1965); *P. plicatulum* (BRYAN & SHAW, 1964)

Pennisetum: review of *P. clandestinum* (MEARS, 1970); the strange case of disjunctive distribution in *P. pedicellatum* (11.5); the hybrid between *P. typhoides* and *P. purpureum* in India (WHYTE, 1964), Queensland (PRITCHARD, 1971) and Uganda (FERNANDES, 1971); apomixis in *P. massaicum* and ten other species (D'CRUZ & REDDY, 1968); seed production of *P. polystachyon* (MISHRA & CHATTERJEE, 1968)

Setaria: Cytology, maintenance of strain purity, polyploid structure and cytology of species hybrids; leading to the review of that closely related group of species within the 125 species of *Setaria*, which are being grown increasingly in the higher rainfall areas of subtropical and tropical Australia, especially the cultivars Nandi and Kazungula, within the so-called *S. sphacelata* complex, widespread naturally in Africa but rarely dominant over large areas (HACKER & JONES, 1969); seasonal yield distribution (HACKER, 1972)

Urochloa: review of species of pasture value (WHITEMAN & GILLARD, 1971); *U. mosambicensis* as companion for *Stylosanthes humilis* in Queensland (GILLARD, 1971); *U. mosambicensis* from dry tropics of Africa, common in overgrazed situations and prolific seed set; tetraploid aposporous apomict (PRITCHARD, 1970).

9.7 C.A.B. Bibliographies

The Commonwealth Bureau of Pastures and Field Crops has published bibliographies on the species of Gramineae listed in Table 9/2 (see also 10.7).

Table 9/2

Species of Gramineae

Species	Number	Number of references	Period covered
Andropogon gayanus	1152	50	1931–1968
Brachiaria brizantha	1093	39	1956–1967
Brachiaria decumbens	1077	12	1954–1967
Brachiaria ruziziensis	1094	29	1956–1967
Cenchrus ciliaris	1127	144	1959–1968
Chloris gayana	1081	193	1958–1967
Cynodon dactylon	1130	350	1959–1968
Panicum maximum	1143	183	1959–1968
Pennisetum clandestinum	934	91	1956–1967
Pennisetum purpureum	1116	200	1958–1968
Setaria sphacelata	1205	106	1957–1969
Sorghum almum	1170	88	1944–1969
Themeda triandra	1119	95	1948–1968
Urochloa spp.	1211	67	1931–1970

10 THE LEGUMINOSAE

10.1 Origin

Botanists appear to agree that the great expansion of the angiosperms in the Cretaceous began in the tropical rain forest (some say on the borders of that zone – TAKHTAJAN, 1969), and that the Leguminosae was one of the early families at that time. CORNER (1949) expounded upon his theory that the fruit of the durian (*Durio zibethinus*, Bombacaceae) was the primitive fruit of modern flowering plants, by showing how this arillate fruit has changed into the dry follicle or capsule with small, often winged, easily detached, ex-arillate seeds, or into the berry, drupe or nut. CORNER found in the Leguminosae eighteen genera with the aril more or less covering the seed, in the sub-families Mimosoideae, Caesalpinioideae, Swartzioideae and Papilionaceae. This is regarded as a relic distribution; nearly every arillate genus shows in different species all stages in reduction or loss of the aril.

TAUBERT (1894) stated that 82 per cent of the genera and 52 per cent of the species in the Leguminosae are tropical. The large number of small genera occurring in tropical regions is a reflection of relic status, e.g. in the most primitive sub-family Caesalpinioideae (TUTIN, 1958).

NORRIS has repeatedly stressed that the tropical legume is the original and typical one, probably evolving under conditions closely approximating those of the modern tropical rain forest, characterized by low soil nutrient status and high acidity. The temperate legume is a derived type (NORRIS, 1966a, 1966b, 1970, 1971); there is good evidence for regarding the ubiquitous, slow-growing tropical 'cowpea type' *Rhizobium* as the archetype from which all other types are derived.

10.2 Taxonomic geography

The principal theme of the Fifth Plenary Meeting of A.E.T.F.A.T. (Association pour l'Etude taxonomique de la Flore d'Afrique tropicale) was the relation of the flora of Africa with that of other continents. BRENAN (1965) contributed on the geographical relations shown by the genera of the family occurring in tropical Africa:

'The genus is a particularly useful unit for studies of this sort since the numbers of species of a given genus occurring in different parts of the world often indicate its centre of origin and directions of spread.

I am of course aware that there are dangers in assuming that the area where the maximum number of species of a genus occur is necessarily its point of origin. Nevertheless this assumption is more often likely to be true than false and where a certain pattern of distribution is found to be repeated by numerous genera, it seems reasonable, unless there is definite evidence against it, to accept the pattern at its face-value and to try to draw conclusions from it.'

Relations are stated between African genera of Mimosoideae, Caesalpinioideae and Papilionoideae and those of South Africa, Europe and the Mediterranean region, Madagascar, India, Asia (the most important link, with no less than 31 genera of tropical African origin showing this pattern of distribution) and America ('spectacular discontinuity'); there are 42 pantropical genera in tropical Africa, seven with their main development in Africa.

The possible centres of origin or at least of wide occurrence in the wild are given in some of the papers discussed below, on those tropical legumes which are of actual or potential economic significance. VERDCOURT (1972) states that the useful properties of tropical *Phaseolus* have scarcely been tapped:

'It is a fact for example that the genus *Phaseolus* (which I have restricted to the relatives of *P. vulgaris* – see *Kew Bulletin* 24: 507 ff 1970)) has not been properly taxonomically revised – a necessary prerequisite for any agronomical work. Even less has been carried out on forage crops. A revision of the Phaseoleae particularly for tropical America and Asia is much needed. The tropical African species are now quite well known and described in the Flora of West Tropical Africa, Flore du Congo Belge and Flora of Tropical East Africa.'

Some of the most promising introductions have been made in Phaseoleae and Hedysareae *sens. lat.* (taking a narrow view of tribes, then particularly the Stylosantheae, to a lesser extent the Desmodieae).

'Further exploration within these groups should be profitable (POLHILL, 1972), especially in South and Central America, which has a rich flora, an exceptional variety of climate and topography and a short history of intensive exploration. Africa has a relatively depauperate flora, but the Leguminosae are quite well represented (1,000 species of Papilionoideae in East Africa alone); there is a good range of climate and elevation, reasonable access for exploration, and quite well-documented taxa' (See reference to plant exploration in Africa by R. J. STRICKLAND, 1972). The potential of Asia in respect of legumes may be more limited.

'The tropical highlands provide a pathway by which tribes such as the Trifolieae, which we presume to have evolved largely in temperate regions, extend into the equatorial zone; in places like East Africa, there is a wide diversity of legume genera from which introductions might well be obtained for other tropical regions' (POLHILL, op. cit; see also 11.5 under *Trifolium*).

Reports of plant collecting expeditions do not always indicate the communities in which a particular legume, for example, has been found. It would seem that most of the legumes already found to have potential value for tropical pastures must grow naturally in relatively mesophytic environments, but where – in association with grasses, or other herbs, and in the open or in forest fringes? A second group are the rugged, unpalatable or even toxic indicators of overgrazing (in India, for example), the species of *Tephrosia*, *Cassia*, *Sesbania* and *Crotalaria*, that are below the grass-dominant communities in the succession towards bare ground. Because of their hardiness, these species also might have a place in the less favourable, more xerophytic environments, if their palatability to livestock could be even partially improved by selection and breeding.

10.3 **Legume/Rhizobium symbiosis**

NORRIS (1970) states the broad outlines of evolution of the symbiosis as follows. As with the Gramineae, development since the Cretaceous has been towards more arid and temperate conditions. Nodulation has been recorded for the first time in leguminous trees of tropical rain forest in Amazonia and Guyana, on fourteen species in nine genera, including three species of *Swartzia* (NORRIS, 1969):

'Evolution by the host legume:

(i) Invasion of temperate and arid zones of the earth
(ii) Reduction of growth form from tall rain forest trees to shrubs and woody lianas to perennial herbs to annual herbs.
(iii) Progressive loss of the ability to establish symbiosis unspecifically with cowpea *Rhizobium*, which is another way of saying progressive development of strain specificity or physiological selectivity for adapted strains of bacteria.

'Evolution by the symbiont bacteria:

(i) Progressive loss of ability to tolerate acid soil conditions where the host legume adapts itself to alkaline soil environments.
(ii) Development of peritrichate flagellation from the primitive condition of a single subpolar flagellum, accompanied by a change from alkali-producing to acid-producing habit in culture media.
(iii) Development of physiological selectivity for the host legume, parallel to that developed by the plant itself. This occurs in two stages, ability to fix nitrogen within the nodule being lost before ability to invade the root and form a nodule.'

Norris continues with reference to cross-inoculation groups: 'The concept of the cross-inoculation group, as originally advanced, supposed that all legumes belonged to one group or other. Those that did not belong to a group were 'exceptions'. In fact, the cross-inoculation group is the exception, not the rule. From the basic unspecialized tropical legume in symbiosis with the cowpea type of *Rhizobium*, every degree of specificity has developed, sometimes in a single isolated species, sometimes in a group of species. The existence of a cross-inoculation group then simply means that we are dealing with a recently developed cluster of species sharing a common degree of physiological specialization.'

There is a considerable literature on the practical techniques of inoculation, with due attention to status of calcium, phosphorus and molybdenum in the soil (Norris, 1967, 1971).

10.4 Fixation and transfer of nitrogen

Three sources of nitrogen are suitable for improving soil fertility in pastures (Henzell, 1968):

(i) soil nitrogen, i.e. nitrogen accumulated in soil organic matter at some time in the past,
(ii) nitrogen fixed by nodulated legumes, and
(iii) fertilizer nitrogen produced by industrial synthesis.

The use of legume-based pastures is now profitable in Queensland, especially with *Stylosanthes humilis* and *Phaseolus atropurpureus* c.v. Siratro. They are important in providing legume as a feed and in changing the feed value of the grass, say, *Heteropogon contortus*, and the weeds with which they are grown. But, in fact, after calculating how much legume nitrogen is available to grasses and weeds (Henzell, *op. cit.*), it is found that: 'fertilizer nitrogen is the only source presently available in Queensland that is capable of maintaining maximum growth rates in tropical grass . . . much of the available nitrogen supply in both permanent pasture or ley farming systems comes from recent additions of animal excreta and plant residues, not from old soil organic matter. However, with many pasture soils, especially those with a relatively low total nitrogen content, addition of readily available forms of nitrogen leads inevitably to a build-up of relatively unavailable nitrogen in soil organic matter. No way has yet been found of preventing or limiting the accumulation of unavailable soil nitrogen, without limiting the supply of mineral nitrogen too . . .. If nitrogen becomes relatively very cheap, what then?

'The answer is likely to depend on how much progress we make in improving our legumes and grasses. Take the legumes. The main long-term research problem is their relatively low yields. This will not affect the commercial exploitation of legumes by farmers and graziers during the next decade or two, but it will be important when there is more emphasis on production per acre in Queensland. There is little doubt that higher legume yields would result in more total growth by the pasture, a higher rate of nitrogen fixation, a better nitrogen supply to the grasses, better quality feed, and more animal production per acre. However, there may be inherent biochemical or physiological barriers in legumes that will hinder selection for higher yields of dry matter. Some very recent research at the University of Queensland indicates that photosynthesis in the tropical pasture grasses is about twice as efficient per unit of leaf area or unit of chlorophyll as it is in the tropical legumes (12.3).

'Yield is not a problem with the grasses, but their low feed value is a major disadvantage. Here, too, there may well be genetic barriers that prevent tropical grasses from being as good a feed as ryegrass, or better still, as good as white clover. The productivity of nitrogen-fertilized tropical grasses would be truly spectacular if the problem of feed value could be overcome.'

Yields would be comparable to those obtained from *Panicum maximum*, *Brachiaria mutica* and other gross-feeding African grasses when cultivated with city sewage water or cowshed wash from dairy colonies to provide the main source of green fodder for milch cattle and buffaloes in India (fed with concentrates to compensate for deficiencies in quality).

Data on the amounts of nitrogen fixed by tropical pasture legumes and transferred to their gramineous companions are given by HENZELL (1962), HENZELL, FERGUS & MARTIN (1966) and JONES, DAVIES & WAITE (1967). *Indigofera spicata, Desmodium uncinatum, Lotononis bainesii, Stylosanthes bojeri* all fix less than *Medicago sativa* or *Trifolium repens*; with adequate superphosphate, 90 to 100 kg. of nitrogen per year is added to the soil plant system in a *Desmodium uncinatum* pasture; eleven legumes were tried with *Paspalum plicatulum* in grazed plots; nitrogen yield of best legume treatment (*Phaseolus atropurpureus*) equivalent to nitrogen yield from grass fertilized with 170 kg. nitrogen as urea per hectare per annum, but dry matter yields equivalent only to those of grass receiving 90 kg. nitrogen per hectare per annum; nitrogen transfer to the associate grass increases with time, greater for *Ph. atropurpureus* than for *Lotononis bainesii.*

10.5 **Interest outside Australia**

The FAO/UNDP project in Panama reports that many of the

leguminous species imported as commercial seed from Australia are indigenous in Central America (HORRELL, 1972); however, the cultivars selected in Australia are susceptible to diseases and pests in Panama, and many are more adapted to the cooler climates of the subtropics. Collection and selection of local ecotypes are necessary, but in the meantime it may be possible to find better adapted material in the Australian collections.

Two cultivars of *Stylosanthes guyanensis* from South Johnstone appear to be disease resistant; *S. humilis* grows well when free of disease, but it completes its life cycle rapidly and decomposes before the end of the rains. *Desmodium intortum* and *Glycine javanica* might be useful in higher parts of the Pacific zone of Panama, *Phaseolus atropurpureus* in the drier parts. *Desmodium ovalifolium* is promising in nursery plots. Species of *Centrosema* are common in pastures, but not highly productive.

In the Caribbean area (SALETTE, 1972) legumes can be of real interest only:
- where rainfall is unfavourable to maximum growth of grasses – less than 1300 to 1500 mm. (*Phaseolus atropurpureus, Glycine javanica*), or
- in inaccessible areas where use of fertilizers is impossible or too costly, or
- in periods when grass growth is poor and legumes may be grown as a catch crop.

Annual low-growing legumes have no place in the permanent grasslands of the Virgin Islands (OAKES, 1970) due to their lack of size and competitive ability, or to the inability of some species to form compatible combinations with grasses. Failure of other species is often attributed to lack of resistance to crop pests and to drought, soil deficiencies and high pH. Low reproductive capacity, lack of competitive ability, liability to damage by trampling and slow rate of recovery are chief causes of failure for many viny forms. However, it is considered that certain legumes are useful in supplying forage and improving its quality at critical times when feed supplies are poor in quantity and quality, but in maximum demand. Unfortunately, standards of livestock and pasture management are poor (see also DE SOUZA, 1966).

Some ten legumes from Queensland are being tested in the llanos of Venezuela (FAO meeting in Cali, Colombia, 1972). BULLER, ARONOVICH, QUINN & BISCHOFF (1970) discuss the performance of tropical legumes in the upland savanna of central Brazil. HYMOWITZ (1971) assembled over a five-year period in the Introduction Garden at Matao a collection of 44 species of indigenous or naturalized Brazilian legumes, together with a collection of symbiotic strains of *Rhizobium*. A source of germ plasm is established of species which show promise as drought-resistant fodder plants, browse and green manure plants:

Stylosanthes gracilis – adapted to low fertility soils, pH 4.5 to 5.5, drought-tolerant
Phaseolus spp. – none of any value
Teramnus uncinatus, *Centrosema pubescens*, *Calopogonium mucunoides* – excellent results
Desmodium barbatum, D. canum, D. uncinatum – poor germination even after scarification; promising at first, but later infected by virus
Leucaena leucocephala – source of protein during dry period, but *Cajanus cajan* less toxic
Arachis – useful new germ plasm for crossing with *A. hypogaea* – *Clitoris, Crotalaria, Desmanthus, Indigofera, Rhynchosia, Sesbania, Tephrosia* – annual and perennial, erect and semi-prostrate shrubs and herbs, perhaps best for green manure (but see 10.2).
Aeschynomene and *Chaetocalyx* – useful browse or pasture in natural stands, but not competitive in improved pasture.

LEME DA ROCHA (1972) reviews work in Brazil (FAO meeting in Cali, Colombia), especially: relation between tropical fodders and beef and milk production; effects of fertilizers on yield and quality of tropical fodders; use of tropical legumes such as *Stylosanthes guyanensis* and *Centrosema pubescens* in association with *Digitaria decumbens* and *Pennisetum purpureum* respectively; studies since 1965 on the legume/*Rhizobium* symbiosis with reference to effects of high soil temperature, manganese toxicity, soil calcium and phosphate. SERPA (1966) discusses the genetical basis of impermeability of the seed coat of *Centrosema pubescens*, and stresses the need to isolate cultivars with permeable seeds.

Populations of *Stylosanthes* have two disadvantages in Côte d'Ivoire (PICARD & FILLONNEAU, 1971). They are affected by the rhythm and height of cutting; production falls after two years. *Centrosema* is higher in nutrient value than *Stylosanthes*, but lower in yield and not worth retaining. Like *Stylosanthes*, this plant does not respond to fertilizers. *Centrosema pubescens* has some value for increasing animal production and improving soil fertility in northern Uganda (STOBBS, 1969c). See bibliography no. 1265 of the Commonwealth Bureau of Pastures and Field Crops for 60 references on nodulation in tropical pasture legumes under African conditions. Growth, morphology and seeding habit of tetraploid Siratro (*Phaseolus atropurpureus*) have been studied in India (SINGH & PATIL, 1970). FAO/UNDP missions (unpublished) recommend the establishment of grass/legume mixtures in the montane zone of Sri Lanka (W. D. ANDREW) and in the hill country of the Philippines (R. E. HARRISON & L. C. SNOOK). E. C. FARINAS has conducted extensive trials with indigenous and exotic legumes in the Philippines.

10.6 **Research in Australia**

Those interested in the history of research on tropical legumes in Australia over the past twenty years may refer to the review articles by BRYAN (1963), DAVIES (J. G.) (1966), and HENZELL (1967), and to HUTTON (1968) in his Farrer Memorial Oration to the Australian Institute of Agricultural Science. Below is given a cross-section of some relevant papers which indicate the direction of current research. Reference should also be made to the Annual Report of the C.S.I.R.O. Division of Tropical Pastures, St. Lucia, Queensland.

There are important climatic regimes in Northern Australia for which no useful legumes are available (HENZELL, 1967). Within the area of climatic adaptation of *Stylosanthes humilis*, there are soils upon which the plant cannot be established easily. Tropical legumes are of limited use in the subtropical areas with less than 890 mm. rainfall per year; little progress has been made in the dry areas of Northern Australia with less than 500 to 550 mm. rainfall per year. Yields of tropical pasture legumes are relatively low compared with tropical grasses receiving nitrogen, as stated above (HENZELL, 1968). . . . 'the problem is to find legumes that grow faster' (JONES, DAVIES & WAITE, 1967).

Effect of temperature on vegetative growth of six tropical legumes has been studied (WHITEMAN, 1968) – *Phaseolus lathyroides*, *Ph. atropurpureus*, *Desmodium uncinatum*, *D. intortum*, *D. sandwicense*, and *Glycine javanica.*

Optimal temperature for growth of all species is 30 to 25 ± 3 °C, lower than values reported for tropical grasses.

Centrosema – yield attributes of species and ecotypes (GROF & HARDING, 1970); interspecific hybridization (GROF, 1970)

Desmodium – review by BRYAN (1969); pasture value (BRYAN, 1966); seasonal changes in growth and nodulation (WHITEMAN & LULHAM, 1970; WHITEMAN 1970a and 1970b); nutritive value and chemical composition (MILFORD, 1967); failure of chromosome pairing (HACKER, 1968)

Dolichos axillaris – susceptibility to frost (JONES, R. M., 1969).

Glycine – chromosome numbers (PRITCHARD & WUTCH, 1964); genotypic variation in growth and nodulation in *G. wightii* (NICHOLAS & HAYDOCK, 1970; NICHOLAS, 1971)

Indigofera – biochemical and nutritional studies in the hepatotoxic amino acid, indospicine (HEGARTY & POUND, 1970; CULVENOR, FOSTER & HEGARTY, 1971; CHRISTIE, DE MUNK, MADSEN & HEGARTY, 1971)

Leucaena – review of research, agronomic, chemical, genetic (GRAY, 1968); rhizobial symbiosis (GATES, 1970); deleterious effects on livestock (DONALDSON, HAMILTON, LAMBOURNE & LITTLE, 1970)

Lotononis bainesii – indigenous to interior of Africa near Tropic of Capricorn, introduced into Australia 1952 (BRYAN, 1961); breeding system and chromosome number (BYTH, 1964); nutritive value and chemical composition (MILFORD, 1967); factors affecting growth (BRYAN, SHARPE & HAYDOCK, 1971)

Medicago sativa – grazing cultivars for subtropics (BRAY, 1970)

Phaseolus – cultivar Siratro produced by crossing two Mexican types of *Ph. atropurpureus* (HUTTON, 1962); flower and seed colours as markers for new derivatives (HUTTON & BEALL, 1971); susceptibility to frost (JONES, 1969); effect of close cutting and nitrogen (JONES, 1967); seasonal changes in growth and nodulation (WHITEMAN & LULHAM, 1970, WHITEMAN, 1970a and 1970b); rhizobial symbiosis (GATES, 1970)

Stylosanthes – (i) assessing agronomic potential (BURT, EDYE, GROF & WILLIAMS, 1970; Northern Territory Agricultural Branch, 1966; WOODS, 1969); computor simulation of yield curves (BYRNE & TOGNETTI, 1969)

(ii) superphosphate, growth and chemical composition (SHAW, GATES & WILSON, 1966); high phosphorus/high sulphur/protein synthesis (GATES, WILSON & SHAW, 1966); superphosphate and annual grasses (WOODS & DANCE, 1970); close relation between concentration of total nitrogen, total sulphur and age (JONES, ROBINSON, HAYDOCK & MEGARRITY, 1971)

(iii) rhizobial symbiosis ('t MANNETJE, 1969; GATES, 1970)

(iv) hardseededness and seed dormancy (CAMERON, 1967e); germination and weather (WINKWORTH, 1969)

(v) photoperiod, flowering, growth habit, dry matter ('t MANNETJE, 1965; CAMERON, 1967a, 1967b and 1967c, 1970).

(vi) chromosome number and morphology (CAMERON, 1967d)

Vigna vexillata – nutritive value and chemical composition in south-east Queensland (MILFORD, 1967)

Teramnus uncinatus – susceptibility to frost (JONES, 1969)

10.7 C.A.B. Bibliographies

The Commonwealth Bureau of Pastures and Field Crops has published bibliographies on the species of Leguminosae listed in Table 10/1 (see also 9.7).

Table 10/1

Species of Leguminosae

Species	Number	Number of references	Period covered
Cajanus cajan	1253	66	1959–1970
Dolichos spp.	1220	20	1960–1969
Lablab niger	1221	62	1960–1969
Lespedeza cuneata	816	12	1950–1966
Phaseolus atropurpureus (syn. *Macroptelium atropurpureum*)	1257	83	1951–1970
Rhynchosia spp.	1234	37	1931–1970
Stizolobium spp.	1222	94	1955–1969
Stylosanthes humilis	1113	69	1931–1968
Trifolium africanum	1290	8	1931–1972
T. burchellianum	1291	17	1931–1972
T. semipilosum	1292	70	1931–1972

11 GENETIC EVOLUTION AND RESOURCES

11.1 Centres of origin

Much of the current discussion of centres of origin, or gene centres, or centres of maximum variability, terms which are not necessarily synonymous, has been concerned with plants on their evolutionary pathway to cultivation and domestication. We find, for instance, that some annual Gramineae became available as human food crops in the early history of man in his ecosystem, generally in biologically critical environments, rather suddenly and over a relatively short time span in the geobotanical history of the plant genera concerned. Some explain this epoch-making event in the history of man in terms of human determinism; others believe that these annual forms of the Gramineae arose from perennials as a result of environmental determinism (WHYTE, 1972b). Acceptance of the latter proposal involves a change from a *centre* of origin to the concept of *zones* between isohyets in sub-desertic environments (HARLAN, 1971).

The situation is different when one tries to define the possible centres of origin of the wild members of the Gramineae, the subfamilies, genera and species. It would seem essential to combine the evidence now becoming available on continental drift with that from pollen core analysis, and on past migrations of plants by natural means and with man's accidental or intentional assistance. Only by such analyses, of which few examples are available, can we eliminate the recent and geobotanically irrelevant, and gradually narrow down the field of claimed distribution to an area or areas which might conceivably be regarded as the place of origin and initial speciation of the genus in question. Renewal of speciation may be assumed perhaps to have occurred in the various stages of geographical migration from that centre; on the other hand, it might not, and the type now far from the centre may show little or no variability (CHAKRAVARTY & KACKAR, 1971, on *Panicum antidotale*; see 11.5).

HARTLEY (1958, 1964) has prepared world distribution maps of the Andropogoneae and the Paniceae. Relative species density is recorded, using the percentage of species of a particular sub-family in the total grass flora in some 300 regions; ARBER (1935) had earlier warned that abundance of species in any particular area could not be taken as indicating that it was a centre of distribution. Floras are to be regarded as unreliable for such studies unless accompanied by an ecological and historical analysis of the total vegetation; they include both true indigenous species (which are usually few) and 'recent' exotics which could not have been constituents of the vegetation before interference by man. One must also know whether the compiler was, in taxonomic argot, a 'splitter' or a 'lumper'.

The Andropogoneae show zones of high concentration in western India and southern Indonesia. India can be accepted, but Indonesia calls for a study of the history of devegetation, fire and shifting cultivation. Although a zone of highest concentration of the Paniceae is found in northeast South America, HARTLEY does not support a New World origin for this sub-family. Rather he suggests a common origin for the two sub-families from a common panicoid stock in the warmer parts of the eastern hemisphere, probably in the East Africa/Madagascar region. This has since been confirmed by BUTZIN (1970a and 1970b), who finds the most 'primitive' representative of the Paniceae, *Pseudolasiacus*, in Madagascar, while most of the genera considered to belong to the same tribe, Microcalamineae, are also African.

11.2 Reconciliation of species taxonomy and variability

CLAYTON (1966) defined the problems facing the taxonomist to the Ninth International Grassland Congress as follows: 'One of the most difficult questions confronting the taxonomist is the concept of species. The species is traditionally defined as a morphologically distinct population; but it is now known that this empirically recognised level of distinctness happens to coincide with a biological phenomenon – the closed breeding system. There are thus two facets to our concept of species, and they are not always easy to reconcile. *Hyparrhenia hirta* (Linn.) Stapf contains at least five different chromosome races which presumably cannot interbreed; but they are morphologically indistinguishable and it is pointless to give names to plants which the majority of users cannot reasonably be expected to identify. In fact ability or failure to hybridize is a test which is often difficult to interpret in taxonomic terms, especially as barriers to interbreeding in nature are often much more subtle than simply inability to set seed after pollination has been effected. Thus *Setaria sphacelata* (Schumach.) Stapf and Hubbard ex M. B. Moss, *S. anceps* Stapf ex Massey and *S. aurea* Hochst. ex A. Br. seem to be inter-fertile, yet somehow they remain distinct in the wild. Apomictic swarms, particularly common in Panicoid genera such as *Paspalum*, defy precise definition on either count, and we can only attempt to recognize units roughly analogous to ordinary sexual species. It is not always appreciated what a wealth of variation lies concealed behind names such as *Hyparrhenia rufa* (Nees) Stapf, *Cenchrus ciliaris*

Linn. or *Panicum maximum* Jacq. Individual specimens seem to come from utterly different species, and it is only in the context of a large collection that they are seen to form part of an amoeboid whole. This is entirely natural for evolution implies that we are bound to encounter emergent species still indistinctly differentiated from the parent population. At the other end of the scale are very narrowly defined species, such as *Digitaria decumbens* Stent, which is probably no more than a single strain selected from *D. pentzii* Stent. Its widespread acceptance as a species disguises the fact that even better forage strains might be available in the parent material. It can be seen that the one rank, species, has to accommodate a vast conglomeration of different kinds of population and is a clumsy jacket in which to encase the subtleties of genetic systems. A more elaborate hierarchy is possible, but the nomenclature then becomes cumbersome and one feels that the present binomials are quite long enough for most purposes'.

11.3 **Significance and use of variability**

If grasses have pioneered into semi-arid and arid environments, or if they have been able to persist in an environment that has become arid, they must have possessed a latent reserve of genetical variability. They must also be able to cope with the different specific spectra which may arise in wet and dry years in the arid zone, or in the drier, falsely grassy areas within a humid zone, where there is wide variability of rainfall between seasons and of distribution within seasons.

LEGRIS (1972) proposes the hydric balance (the product of the xerothermic index and deficit of saturation) for the appreciation of the amplitude of bioclimatic variations that would have resulted in the palaeoclimatic substitution of one type of vegetation by another. In other words, these bioclimatic variations would have been sufficient to permit the migration of a flora from one region to another, or would have created conditions of physiological stress in the environment of a fixed plant population, leading to chromosomal aberrations and speciation (12.8).

MEHER-HOMJI (1964) discusses life forms and biological spectra as epharmonic criteria of aridity and humidity in the tropics. A successful grass species will always have types available within a natural population to cope with conditions in whatever fluctuating ecological niche they may find themselves. The lower the rainfall, the wider the annual variations in its amount and geographical and seasonal distribution and so the greater the need for variability in grass species.

If maximum variability in the grasses indicates proximity to their centre of origin, it would be desirable for an international cooperative study to examine the degree of variability within one species over its whole range of distribution as defined by the taxonomists. Conversely, it must be decided whether rarity of a species is due to its high palatability, which exposes it to those biotic factors which drastically reduce its frequency in the plant communities.

Relevant to the proposition that western Monsoon Asia represents a centre of high variability in certain tribes of the Gramineae, and therefore a centre of origin for the Andropogoneae (HARTLEY, 1958, 1964) is the work of OKE (1968) on the grasses of Maharashtra. Table 11/1 shows his analysis of biotypes in six of the important genera in that part of the Deccan (see also OKE, 1971).

Table 11/1

Numbers of species and biotypes within certain genera of the Andropogoneae (OKE, 1968)

	Number of Species in world	Species in India before analyses by J.G. OKE	Species recognized	Biotypes	Species analyzed for biotypes
Ischaemum	50	35	15	41	in 14 species
Sehima	6–9	5	5	8	*S. nervosum* and *S. sulcatum*
Chrysopogon	18	8	8	2	*Ch. montanus*
Dichanthium	8	7	6	36	*D. annulatum* *D. caricosum* *D. aristatum*
Heteropogon	8	6	6	3	*H. contortus*
Iseilema	6	6	6	5	*I. wightii, I. laxum*

11.4 The Dichanthium complex

CHAPMAN (1970) quotes a personal communication from J. R. HARLAN that, on the basis of cytological analysis, the three taxa which were formerly considered sufficiently distinct to merit generic status – *Bothriochloa*, *Capillipedium* and *Dichanthium* – should now more reasonably be regarded as one genus under the prior name *Dichanthium*. Within this genus, the cytogeneticists claim that they can distinguish four kinds of species (HARLAN, 1966, not seen and therefore quoted from CHAPMAN, 1970):

endemic Indian diploids; old relicts, 2n = 20

Australian endemics; diploids and polyploids

American species; high polyploids formerly grouped under *Bothriochloa* 2n = 60 to 2n = 220

Afro-Asian weed species; wide-ranging, 2n = 20, 40, 50, 60 or 80 Eurasia, north, east and south Africa, south-east Asia (see 9.5.2) and across to the South Pacific to Australia; only the diploids are sexual; the tetraploids are facultative apomicts reproducing sexually at low frequency; the higher polyploids are virtually obligate apomicts (12.8).

The cytogenetical conclusion appears to be (CELARIER, MEHRA & WULF, 1958; MEHRA, 1961, 1962; MEHRA & SINGH, 1968) that the *Dichanthium annulatum* complex shows four distinct ecotypes, tropics, Mediterranean, Senegal and South Africa.

Cytological studies are made in laboratories and greenhouses; geographical interpretations are evolved in the hope that they may fit. Let us try the opposite approach, starting with a simple, ecological/historical proposition, and then ask the cytogeneticists whether the results of their more complex studies fit. The interpretations of cytogeneticists and taxonomists cannot be fully accepted until they have been submitted to the discipline of ecological/historical analyses.

Let us suppose that ancestral Indian and Australian species of this so-called complex were cohabitors at higher elevations in the Mesozoic tropical rain forest environment of Gondwanaland (AXELROD, 1970) before India and Australia parted company, and before what is now called the Indian subcontinent crossed the equator to join southern Asia. Species of the genera *Dichanthium* and *Bothriochloa*, separate or in a taxonomic complex, are important and indigenous members of the grass covers of the subcontinent (DABADGHAO & SHANKARNARAYAN, 1972), probably ancient from the days before man's interference. Being inveterate nomads like *Heteropogon contortus*, *Cynodon dactylon* and *Imperata cylindrica*, they tried to advance into all manner of environments within the subcontinent and into the Mediterranean area (carried as the feed for Arab horses) where they faced physiological stress. The responses to that stress were increasing ploidy plus different intensities of sterility and failure of meiosis in the androecia and gynoecia, which are the result of floral malnutrition and which the cytogeneticists call facultative or obligate apomixis (12.8), if and when these characters become genetically fixed.

Indian traders were active in East Africa long before the arrival of the Europeans, and would have been the medium for the accidental introduction of Indian grasses into the coastal lands and uplands of East Africa. These foreigners (exotics) may have become established in the grass covers of the savannas of the forêts claires, but more probably as ruderals on abandoned cultivated land. When the slave ships sailed from Mombasa and Zanzibar for the New World with slaves brought down from the interior, their bedding would have been grass hay and their staple food mealies, with their weed seeds. The East African bedding was probably changed at the Cape and replaced with clean grass hay brought down from the grassveld or the cultivated areas.

Thus the New World may at various periods in history have received non-indigenous grass seeds, with the introduction of Zebu cattle, with the African slaves, and in the scrips and fardels of the latterday plant collectors; in this entirely new environment and in the face of the negligible competition from the resident American species, this new material exhibited that massive explosion of renewed speciation and polyploidy which is the pride and joy of the taxonomist and cytogeneticist. But these exotics are irrelevant in the geobotanical history of the New World; nowhere are they truly indigenous members of the ancient grass covers which existed before man.

11.5 **Reproduction and variability in species**

Bothriochloa pertusa – cultivar from pure line selection (MAHARASHTRA STATE, 1969; OKE, 1971).

Brachiaria decumbens – until sexual forms of this obligate aposporous apomict have been found, it will not be possible to produce improved varieties by conventional breeding techniques (PRITCHARD, 1967).

Brachiaria mutica – first introduced into India at Quilon, Kerala; now highly important feed for dairy industry in southern part of the country; genetical base requires consideration.

Cenchrus – distribution according to BOR (1968): Southern Europe (Sicily), Syria, Israel, Sinai, Egypt, Arabia, Kuwait, Bahrain, Iran, Afghanistan, India, north Africa (Morocco, Algeria, Tunisia, Libya), Micronesia, tropical Africa, south Africa, introduced into Australia and North America;

– working at the Indian Agricultural Research Institute, New Delhi, B. D. PATIL, (PATIL & SINGH, 1963; PATIL, SINGH & GHOSH, 1964) isolated some 53 distinct types among a collection made in western India of species of *Cenchrus* which are in different reports called *C. ciliaris* or *C. setigerus*; the character of possible value for pioneering into habitats with contrasting rainfall regimes is 'season of maturation', i.e. growth active

(i) throughout the year, or in
(ii) spring, rainy season and winter
(iii) summer, winter and spring
(iv) spring and summer, or
(v) summer

– working at Central Arid Zone Research Institute, Jodhpur, Rajasthan, CHAKRAVARTY and others have studied polymorphism, variation in seed yielding characters, and variation in morphological and physiological characters in *C. ciliaris* (CHAKRAVARTY & DAS, 1965; CHAKRAVARTY & KALKANI, 1966; CHAKRAVARTY, RATAN & MURARI, 1970); also variation in *C. setigerus* (CHAKRAVARTY & BHATI, 1968);

– hybridization of sexual *C. ciliaris* and *C. setigerus* at College Station,

Texas, has resulted in 'a vast array of sexual and apomictic types, some of which might easily be classified as new species if discovered in a natural habitat' (BASHAW, HOVIN & HOLT, 1970);
– population differences in *C. ciliaris* in Gujarat (PANDEYA & JAYAN, 1970);
– salinity response of cultivars of *C. ciliaris* (GRAHAM & HUMPHREYS, 1970).

Chloris gayana – reported to reproduce sexually in Africa and by apomixis in Australia; tetraploid race an aposporous apomict; cytological study of two diploids and five tetraploids in Queensland confirmed a normal sexual method at both ploidy levels (JONES & PRITCHARD, 1971).

Cynodon – for the work of HARLAN, DE WET & RAWAL (1970a and 1970b) see 5.6.
– *C. dactylon* highly variable species, ranging from low, fine-leafed turf types to coarse-stemmed varieties over 1.2 m. tall; high degree of heterozygosity; highly cross-pollinated; *C. transvaalensis* much more uniform but still variable enough to suggest that it is cross-pollinated and reproduces sexually; breeder wishing to improve species should collect a large group of variable types; if still not enough, variability may be created by somatic mutations, intra- and interspecific hybridization (BURTON, 1966);
– *C. polystachyus* – natural fertility and chromosome numbers (OKIGBO & CHHEDDA, 1966).

Desmodium intortum – opportunity to develop fairly large gene pool for use in different environments, provided material is accurately catalogued or indexed according to adaptation (ROTAR, 1970);

Dichanthium annulatum – distribution according to BOR (1968): – widespread in India and Burma, tropical and north Africa, not so far found in Sri Lanka; see 11.4 and PANDEYA (1953); cultivar from pure line selection (MAHARASHTRA STATE, 1969, OKE, 1971)

D. caricosum – cultivar spreading type, from pure-line selection (MAHARASHTRA STATE, 1969; OKE, 1971).

Digitaria decumbens/pentzii – first introduced into Florida as a few cuttings, possibly from the same plant; subsequent collecting expedition in Africa has widened dangerously narrow genetic base (OAKES, 1965).

Heteropogon contortus – HACKEL's subspecific classification not accepted (TOTHILL, 1968); two scales of variation recognized: (1) primary scale closely correlated with a polyploid series, accounts for a general pattern of morphology associated with broad geographic distribution, and (2) secondary scale concerns patterns of phenologic variation within geographic regions, which seems to be largely associated with environmental variations; apomixis is suggested as an important process in developing and maintaining the pattern of variation within this scale.

'It is considered that occasional reversion to sexuality is responsible for generating the variability which is observed in populations of *H. contortus* . . . only occasional sexual forms need arise, which, by establishing in new niches during times of abnormal environmental change or migration of the species into new areas, would generate great arrays of new biotypes, both apomictic and sexual. So the cycle would continue and a new equilibrium establish itself' (TOTHILL, *op. cit.*).

Variations in flowering behaviour and relative male: female flower ratio in the polygamous inflorescence of this obligate apomict are controlled by daylength; this gives some hope of manipulating apomixis to obtain some sexual recombinations (TOTHILL, 1970).

Hyparrhenia rufa – variability for growth type and flowering characters (CUANY, 1967).

Iseilema wightii – cultivar from pure-line selection (MAHARASHTRA STATE, 1969; OKE, 1971).

Lasiurus sindicus – variation in morphological and physiological characters (CHAKRAVARTY & KACKAR, 1970).

Panicum – an attempt has been made at the Indian Agricultural Research Institute, New Delhi, to build up a cytotaxonomic classification, by integrating data on morphological characters, features of epidermal pattern, chromosome number and behaviour during meiosis, and pollen characters. The presence of primitive characters in *P. antidotale* seems to confirm the earlier views of other workers regarding an evolutionary relation between the Panicoideae and the Bambuseae (JAUHAR & JOSHI, 1966; other references in WHYTE, 1973). Other studies have indicated a probable evolutionary history for *Panicum maximum* (JAUHAR & JOSHI, 1969; but see below under ORSTOM).

Panicum antidotale – exhibits little variation in the natural vegetation in Rajasthan, apart from plant height, number of tillers, basal diameter and herbage yield in six plants which may provide useful germ plasm (CHAKRAVARTY & KACKAR, 1971).

Panicum maximum – abundant in humid forest zone of lower Côte d'Ivoire, (rainfall not exceeding 2,200 mm. and not less than 1,200 mm.) on the borders of forests and along maintained roads; scarcely penetrates the deeply shaded slight slopes, where it may rarely be represented by small islands on favourable sites such as areas cleared of forest and brush, or around a temporary water source; with progressive deforestation the species is advancing northwards, into land affected by human activities; in the zone north of the forest, the species is absent in the savanna itself, only very small populations being found in humid hollows at the edge of small forest galleries or along the banks of small streams.

– in his collecting mission to Africa 1971–72, STRICKLAND (1972) collected types of *Panicum maximum* from areas with rainfalls varying between 1,600 to 250 mm per annum. It was noted that this species was invariably associated with woodland in the better rainfall areas, where it assumed the robust form familiar to specialists in fodder cultivation. 'In drier regions of Mozambique, Rhodesia, Botswanaland, Zambia and South Africa, however, it is only to be found around the base of trees growing in shade. On the Kalahari Sands of Rhodesia and Botswanaland it behaves as a slender annual around the base of trees, and more rarely as a biennial.'

– specialists of the Centre ORSTOM (Office de la Recherche scientifique technique Outre-Mer) at Adiopodoumé, Côte d'Ivoire, have studied the relation between systems of multiplication and distribution and phenotypic variability (COMBES & PERNES, 1970; PERNES & COMBES, 1970); three analyses: cytological study of chromosome numbers and embryo sacs, observation of progenies of self-pollinated seeds (sexuality 2 per cent) and of cuttings or tillers, and statistical analyses of differentiation of natural populations.

– the *P. maximum* complex is formed from a type n = 8, origin probably in the diploid populations of type found in Kenya; hypothesis of JAUHAR & JOSHI (1966) deriving *P. maximum* from species 2n = 36 by loss of four chromosomes should, according to ORSTOM specialists, be rejected.

– multiplication by cuttings or tillers produces more variation than multiplication by apomictic seeds, but only as *quantitative* variation within the same *qualitative* phenotype; the place of origin within Côte d'Ivoire of the dominant qualitative phenotype II and some of the quantitative characters are closely linked; apomixis prevents qualitative variation from occurring in 98 per cent of the progeny.

Pennisetum – research at the Indian Agricultural Research Institute, New Delhi, on genome relations and the phyletically basic chromosome numbers shows that it is a polybasic genus comprising several species with chromosome numbers as multiples of 5, 7, 8 and 9 (JAUHAR, 1968; other references in WHYTE, 1973); it has been inferred that *P. typhoides* and *P. purpureum* are phylogenetically related; this must have occurred in Africa, as these species are only doubtfully indigenous (in the historical sense) in the Indian subcontinent.

Pennisetum pedicellatum – an annual of the Paniceae that occurs in West Africa and India; BOR (1960) states that 'despite this extraordinary discontinuous distribution, the specimens from both areas seem to be identical.'

– CHATTERJEE & KUMAR (1964) recognize 25 morphologically different biotypes among a collection of local types from Bihar and of introductions, including a perennial from Australia (CSIRO 7968). B. D. PATIL & R. GHOSH (unpublished) recognize 34 distinct types among

45 accessions from India, Africa and Australia (see also SINGH & ARORA, 1970).
– BARRAULT (1971) reports the collection of 79 ecotypes in regions in the Cameroons with between 600 and 1.200 mm. rainfall. These ecotypes could be grouped in four classes on the basis of cycle from sowing to ear formation, each adapted to a particular environment. The species is the basis of fallow in north Cameroons; beyond and above the 1,200 mm. isohyet, it is replaced by perennial forms of *Andropogon* and *Hyparrhenia*.
Following the proposition made in Chapter 9.4.3, an attempt should be made to recreate this annual from its nearest perennial relative, perhaps a depauperate, drought-affected type of *Pennisetum purpureum*. Such a source of origin would explain its simultaneous origin at two places so widely separate.
Pennisetum purpureum – distribution according to BOR (1960): native of tropical Africa, but now introduced into many other tropical countries;
– this grass and its hybrid with *P. typhoides* are important sources of feed for dairy husbandry in India; history of introduction, genetical base of the species in India compared with probable centre of origin in West Africa calls for review; selection from hybrid derivatives (GROF, 1969); chemical composition of leaves and stems of 21 cultivars in Venezuela (RODRIGUEZ & BLANCO, 1970).

Trifolium – the species of *Trifolium* of Africa south of the Sahara (10.2) were little known until agronomists in Africa and Australia found that a number were potential pasture plants; information is available on cytology, rhizobium affinities and breeding systems (literature references in PRITCHARD & 't MANNETJE, 1967; 't MANNETJE & PRITCHARD, 1968); the effects of photoperiod and night temperature, and the breeding systems of 12 species and varieties; one annual and the three perennial species are allogamous and self-incompatible; eight annuals autogamous – it has always seemed strange to the writer that no one has yet found a use for the self-seeding winter annual (wet monsoon-avoiding) types of *Trifolium repens*, evolved from Louisiana White, and now naturalized in plots in Indian Agricultural Research Institute, New Delhi, and the Fodder Research Station, Haringhata, West Bengal.

11.6 Genetic resources

11.6.1 OBJECTIVES

It is possible to refer only briefly to some aspects of this broad subject as it relates to the maintenance, collection and utilization of the grasses, legumes and other plants for economic use (SUNDARARAJ & MEENAKSHISUNDARAM, 1955; EVANS, 1967; BURT & HAYDOCK, 1968; BURT & EBERSOHN, 1969; MCWILLIAM, 1969; see also FAO Plant Genetic Resources Newsletter, CSIRO Plant Introduction Review and ICAR/IARI Plant Introduction Reporter).

These plants may be required for cultivation as crops on arable land (cultivated pastures, in rotation or otherwise, and fodder crops). For that purpose, it may be correct, as some have suggested (WEBB, 1966), to collect plants on the forest fringes, before assessing their performance in plant introduction nurseries, and passing them on to the breeder, who will change them into 'crop plants'. Most of the important pasture grasses of the world have been derived from restricted habitats such as forest fringes rather than from the great grassland regions of the world (HARTLEY & WILLIAMS, 1956).

The genetic resources from the savannas and tree/grass ranges may also be required increasingly for range reseeding projects, especially where a good cover of perennials has been reduced to a poor perennial or annual status. Under such conditions, normal ecological progression would be too slow, or impossible in the absence of seed mother plants; surface seeding or sowing after ploughing becomes justified, especially if it can later be combined with some degree of livestock control or fencing. For such conditions, the species of the forest fringe would be too tender and mesophytic; one needs to collect seeds from an open stand of a species known to be higher in succession on the degraded site to be treated.

11.6.2 CONSERVATION OF VEGETATION FOR GENETIC PURPOSES

There is much talk of the need to establish areas of natural vegetation in which it is desirable that some potentially valuable and threatened genetic resource should be preserved for future use by collectors and plant breeders. This is not a particularly simple objective to achieve. Perhaps it is unusually difficult in the case of herbage grasses and legumes. In most areas, full protection would involve progression to a higher form of vegetative cover, a marked change in the generic and specific composition of the grass cover, leading ultimately to the elimination of gramineous constituents when the regenerating forest canopy closes over. Protection in such cases is not a practicable proposition without some type of controlled management. Yet conservation of these resources is just as important as with any other economic plants, since the quality of the resource is being rapidly eroded away, nutritionally poorer species taking the place of superior species, xerophytic ecotypes replacing the more mesophytic ones within the species. As with so many crop plants, the field of choice for the collector is becoming progressively narrower.

11.6.3 Technique of Economic Exploration

Selection by the plant collector of geographical areas and of particular grass communities should be made within the terms of the historical context. The breakdown of stated global or regional distributions of species into centres of origin and/or maximum variability and those fringe areas of recent introduction and geobotanical irrelevancy will help to pinpoint areas likely to be most profitable for collection. It is necessary to distinguish between derived savannas and true ancient or indigenous savannas: the former are poor, the latter rich in floral composition, and probably also in continuing variability leading to speciation. Groups of species, with a limited range of ecotypes and genotypes within the species, may demonstrate renewed speciation in quite new directions, following mass migration over centuries from one environment to another rather different one.

The would-be conserver and collector of the genetic resources of the wild Gramineae needs to be fully aware of the historical status of the material in which he is interested. Every species has to be considered independently. Are the species belonging to Indian and African genera now to be found in south-east Asia and eastern Monsoon Asia near the limits of their distribution from the centre, and do they therefore present a limited range or spectrum of variability of little value to the plant breeder? Or has their exposure to new environmental conditions induced genetic change approaching speciation, and thus made them of renewed interest and potential for the plant breeder? (Whyte, 1972b).

Many of the major genera in the Indian subcontinent have come from Africa by migration and/or human agency (*Panicum*, *Digitaria*, *Pennisetum*, *Brachiaria*, *Aristida*, *Eragrostis*). Again, should the plant collector concentrate on their places of origin in Africa, or also on their areas of distribution in the Indian subcontinent, supposing that the change to the more biologically difficult monsoonal environment has induced, in non-apomicts, renewed genetical change approaching speciation, following exposure to physiological stress? In the genus *Panicum*, for instance, Bor (1960) accepts 32 species but discards 375. This would seem to indicate that the taxonomists are dealing with a genus in an advanced state of progress towards renewed speciation in this new environment.

It is important to know the primary objective of the plant collector; is he more concerned with the apparent economic superiority of the material that he is collecting, or does he wish to ensure that he obtains a sample which will demonstrate the full range of variability? The latter would appear to be the more desirable objective, provided time and facilities permit. A knowledge of the taxonomic geography of a species is essential in the planning stages of collection; this provides guidance for field cross-sections designed to tap the full ecotypical and growth-form range within the species to be collected. Preliminary multiplication in or near the environment of collection will ensure that the complete range of variability may be expressed.

Collection need not always be planned in climatically homologous areas. British plant breeders have shown particular interest in collections from the Mediterranean environment. Plants from high altitudes and low latitudes would combine frost-resistance and adaptation to short days. ATKINSON (1970) collected legumes and grasses along several altitude clines (sea level to 4,000 m.) at different latitudes in Mexico, Central and South America. These are being tested as potential winter crops in temperate southern Australia.

11.6.4 Gene Banks

It is debatable whether cross-fertilized herbage grasses and legumes can be maintained on a long-term basis in an acceptable genetic condition in plant introduction nurseries or gene banks. No doubt, the specialized technique necessary will evolve with experience obtained at the Indian Agricultural Research Institute, New Delhi and substations, CSIRO, at Canberra, Brisbane and subcentres; Central America (LÉON, 1971), Matao, Brazil (HYMOWITZ, 1971), and Colombia (GROF, 1972); the last being a proposal of the C.I.A.T. (Centro Internacional de Agricultura Tropical) to collect grasses and especially legumes from South and Central America, and to maintain desirable material, after assessment of potential, in a germ plasm bank.

12 PLANTS AND ENVIRONMENT

12.1 Autecology

Having considered the types of vegetation on tropical grazing lands, the communities, constituent families, genera and species, we turn to the individual plant, its reaction to the environment to which it is exposed, and its requirements from that environment for normal growth and development.

The objectives and terminology of the range manager and fodder agronomist have to be translated into their scientific equivalents in order to show their relations with the fundamental sciences. The effects of grazing and cutting become the physiological response (especially carbohydrate reserves) of the leaves, stem and roots to repeated defoliation (see review article by BOOYSEN, TAINTON & SCOTT, 1963, and subsequent series of studies on *Themeda triandra*, *Hyparrhenia hirta*, *Cymbopogon excavatus*, *Eragrostis curvula*; last in series, STEINKE & BOOYSEN, 1968; RETHMAN & BOOYSEN, 1968a; SANTHIRASEGARAM, COALDRAKE & SALIH, 1966; TARDIN, CALLES & GOMIDE, 1971). The effect of fertilizer is plant nutrition and metabolism (Table 12.1). Nutritive value is the biochemistry of quality, palatability and intake (Table 12.2). Seed production is the result of the processes of floral initiation, flowering and the maturation of zygotes which are governed by light and temperature in correct combination and sequence. Failure to produce seed or to yield cytogenetically normal progeny is evidence of physiological stress caused by undernutrition, malnutrition or lack of turgour due to water deficiency in the flower.

Most of the literature concerns the relation between domesticated grasses, i.e. cultivars that are no longer capable of returning to the wild, rather than the species of naturally occurring communities which are the main concern in this context. COOPER (1970b) has stated: 'Crop production consists basically of the conversion of solar energy and soil nutrients into economic end-products which may be either human or animal foodstuffs or industrial raw materials. The basic limitation to production is, therefore, the seasonal energy input from the sun, but the ability of the crop to make use of this energy is often limited by other climatic factors, such as low temperature or water stress, or the availability of soil nutrients.

'From a knowledge of the environmental limits to production in a particular locality, and the present level of crop production, it should be

Table 12/1

Nutrition

Species	Nutrient	Country	References
Legumes	General	Australia	ANDREW, 1965
Grasses	Nitrogen	Australia	COLMAN & LAZENBY, 1970
Grasses	Nitrogen	Australia	GARTNER, 1969a and b. GARTNER & EVERETT, 1970
Grasses	Phosphorus	Hawaii	CAMPBELL, SHERROD & ISHIZAKI, 1971
Legumes	Phosphorus	Australia	ANDREW & ROBINS, 1969a
Legumes	Potassium	Australia	ANDREW & ROBINS, 1969b
Grasses	Potassium	S. America	FERNANDES, GOMIDE & BRAGA, 1970
Legumes	Copper	Australia	ANDREW, 1963
Legumes	Manganese	Australia	ANDREW & PIETERS, 1970a
Legumes	Salt	Australia	HUTTON, 1971b
Brachiaria brizantha	Nitrogen	Sri Lanka	APPADURAI & ARASARAT-NAM, 1969
Brachiaria mutica	Nitrogen	India	RAI, SINGH & SINGH, 1968
Cenchrus ciliaris	General	Ghana	ASARE, 1970
Chloris gayana	Nitrogen	Australia	HENZELL, MARTIN & ROSS, 1970
Digitaria decumbens	Nitrogen	Venezuela	RODRIGUEZ, 1970
Setaria sphacelata	Nitrogen	Australia	HACKER & JONES, 1971

possible to assess the relative potentialities of either modifying the crop environment, or selecting better adapted varieties, or both. Certain environmental limitations can be avoided by such measures as irrigation or increased fertilizer input, including the application of trace elements. On the other hand, such climatic features as the seasonal light and temperature regime, including photoperiod, and seasonal precipitation cannot be modified by the farmer and the only solution may be selection of adapted genetic material often with appropriate life cycles which synchronise with the seasonal climatic inputs.

'In assessing the potential production of a crop in a particular region, we need to know (i) the environmental limitations to production, both climatic and edaphic, and (ii) the extent to which crop varieties can be developed which are adapted to these environmental limitations. This, in its turn, involves an assessment of how far adaptation, and hence potential production, can be increased by the introduction of new genetic material.'

BREESE (1969) states that the occurrence of genotype/environment reactions has long provided a major challenge in obtaining a fuller understanding of the genetic control of variability. Recent studies have shown that genetic expression can be simply and predictably related

Table 12/2

Chemical composition and nutritive value and intake

Plants	Scope of study	Reference
Grasses and legumes	chemical composition	ABIUSSO & SCHULZ, 1970
	chemical composition and fodder value	ABIUSSO, 1970
Grasses	nutritive value	THORNTON, LONG & MARSHALL, 1968
Grasses	protein and phosphorus content	CHAKRAVARTY, 1968
Grasses and legumes	water-soluble carbohydrates	HUNTER, MCINTYRE & MCILROY, 1970
Grasses	dry matter, leaf angle, silica content	LUDLOW, 1969
Grasses and legumes	lactic acid in silage fermentation	CATCHPOOLE, 1970
Grasses and legumes	protein content and nutritive value	MILFORD & HAYDOCK, 1965
Cenchrus ciliaris	crude protein/crude fibre /N-free extract	TAERUM 1970c
Chloris gayana	crude protein/crude fibre /N-free extract	TAERUM 1970c
Chloris gayana	digestibility and climate	MINSON & MCLEOD, 1970
Digitaria decumbens	N content, leaf/stem ratio; environment	SALETTE & DUMAS, 1970
Digitaria decumbens	crude protein	THOMAS & MCLAREN, 1971
Digitaria decumbens	chemical composition of hay	CHICCO, RODRIGUEZ & FUENMAYOR, 1971
Digitaria decumbens	energy values and nutritive value indices	MINSON & MILFORD, 1966
Eragrostis superba	crude protein/crude fibre /N-free extract	TAERUM, 1970c
Heteropogon contortus	polysaccharides/xylan	BLAKE & RICHARDS, 1970
Panicum maximum	crude protein/crude fibre/N-free extract	TAERUM, 1970c
Pennisetum purpureum	crude protein/crude fibre	BALACHANDRAN, 1969
Pennisetum purpureum	times of cutting/chemical composition/ three cultivars	MARQUES VIERRA & GOMIDE, 1968
Pennisetum purpureum	chemical composition 21 cultivars	RODRIGUEZ & DE BLANCO, 1970
Phaseolus atropurpureus	energy values and nutritive value indices	MINSON & MILFORD, 1966
Salix humboldtiana	protein/energy values/silica/calcium/ phosphorus/magnesium	ABIUSSO, 1969
Setaria spp.	digestibility and climate	MINSON & MCLEOD, 1970
Setaria sphacelata	oxalate poisoning	JONES, SEAWRIGHT & LITTLE, 1970
Setaria sphacelata (*anceps*), *S. splendida* and *S. sphacelata*	digestibility of leaves, leaf sheaths and stems	HACKER, 1971
Themeda triandra	crude protein/crude fibre/N-free extract	TAERUM, 1970c
Tripsacum laxum	dry matter/crude protein/cellulose/soluble carbohydrate	GOMIDE & TARDIN, 1969

to the environment when the latter is measured by its effect on the character under study. Breese discusses results obtained 'when these methods are applied to yield data in herbage plants, which perhaps of all cultivated crops have to yield under the greatest diversity of climatic, edaphic and management conditions.

'They are grown under very different soil conditions, with varying managements, and are normally successively harvested throughout seasons and over years. Thus an appropriate dynamic representation of genotypic reaction to varying environments is of the utmost practical importance as well as of theoretical interest'.

12.2 Potential production and energy conversion

In his review of the literature up to 1969 (106 references), COOPER (1970a) refers to an earlier review (COOPER & TAINTON, 1968) on the light and temperature requirements for the growth of young vegetative material of temperate and tropical grasses, and the contrasting seasonal energy inputs and temperature regimes of temperate and tropical environments (STEWART, 1970). COOPER (1970a) examines the efficiency of energy conversion through photosynthesis, and hence potential dry-matter production at three levels of increasing complexity:

(i) the individual leaf or young seedling grown without mutual shading,
(ii) the closed crop canopy which is intercepting all the incoming light, and
(iii) the established crop grown over a period of several months or years and subjected to regular grazing or cutting.

KAWANABE & HIRAKAWA (1971) have also compared tropical and temperate subfamilies (Panicoideae, Eragrosteae and Festucoideae) and genera of herbage plants in respect of their morphology, behaviour, ecology and distribution, particularly in Japan, the Ryukyus and Taiwan. Criteria included in this review are response to day and night temperatures, amino-acid content and total protein of seed, net assimilation rates and dry matter production, light intensity and leaf area index, and daily and seasonal variations in growth rates and productivity.

For the tropical species such as *Panicum maximum*, *Pennisetum purpureum* and *Brachiaria mutica*, gross-feeding grasses of African origin for which it will become increasingly difficult to find land on which their massive requirements of water and soil nutrients can be provided, a potential production of over 40 tons green per hectare is feasible. Humid tropical areas have the greatest potential production in climatic terms, provided sufficient water, sewage effluent or cowshed wash are available. In subtropical regions, despite a high energy input, production of tropical species is likely to be limited by low-winter temperatures. It then becomes a question of rotating a rapidly growing, high-yielding tropical annual with a temperate annual which will produce where winter conditions are not too severe to prevent growth, e.g. in northern India and Pakistan, and in Kyushu in Japan.

But over most of the intertropical zone, species of these potentialities cannot be grown because of drought. At the extreme, one notes the observation of GAFF (1971) on the unique abundance of higher plants in the southern African flora which can withstand virtually complete desiccation. Four grasses are found to possess desiccation-tolerant mature foliage, able to endure dehydration in the laboratory to 30 to 40 per cent relative humidity; *Oropetium capense*, *Sporobolus stapfianus*, *Eragrostis denuda* and *Microchloa caffra*. Agrostologists working along the southern fringes of the Sahara, or in Pakistan and western India, would be able to add more species to this list of maximum tolerance of drought. Physiologists may find that such species are more efficient energy converters in relation to water intake than the perhaps wasteful species of *Panicum* and *Pennisetum* mentioned above.

The preliminary results of an I.B.P. project for modelling the primary productivity of the world in relation to environmental conditions have been presented by LIETH (1972); the data for grasslands, tropical and temperate, are extracted in Table 12/3.

Table 12/3

Estimate of net primary productivity and energy fixation for the world's grasslands around 1950 (from LIETH, 1972)

Vegetation unit	Size 10^6 km^2	Net primary productivity Range g.m.$^{-2}$we.y^{-1}	Approx. mean	Total for area 10^9t	Combustion value Kcal/g	Annual energy fixation 10^6cal/m^2	10^{18} cal. per area (Col. 2)
Grassland	24.0		600	15.0			60.0
Tropical grassland	15.0	200–2000	700	10.5	4.0	2.8	42.0
Temperate grassland	9.0	100–1500	500	4.5	4.0	2.0	18.0

This incredibly theoretical exercise will be of little value to the grassland practitioner, or to the surveyor of land resources for the production of protein foods for tropical peoples.

12.3 **Photosynthesis**

Studies on growth and photosynthesis have been concerned with tropical species per se, with the differences between tropical and temperate species, and with an attempt to isolate temperate and tropical types from within the same genus. Several methods have been evolved to determine the photosynthetic productivity of a plant cover, especially that of measuring the quantity of CO_2 absorbed by the leaf or by the plant cover. MONTENY (1971b) applied the currently unpopular method of measuring the increase of dry matter per unit area of leaf surface, twice daily on plants in the field, from one hectare of a homogeneous clone of *Panicum maximum* reproduced vegetatively and growing in natural and variable conditions of light, temperature and CO_2 concentration. This gravimetric method gives a useful indication regarding the photosynthetic capacity of a plant cover as a function of the climatic data.

CONNOR & CARTLEDGE (1970) compared observed and calculated photosynthetic rates of communities of *Chloris gayana.* WEST, BIGGS & BASKIN (1968) studied growth and photosynthesis of *Digitaria decumbens* in a gradient of temperatures. LUDLOW & WILSON have published a series from the Botany Department of the University of Queensland (most recent, 1971a and 1971b). In these last papers are described (a) the effects of current environmental factors on leaf net photosynthetic rate and carbon dioxide transfer resistances of plants which had a common environmental history, and (b) how the temperature and illuminance at which leaves develop and to which they are exposed prior to measurement influence photosynthetic characteristics measured under standard conditions. Tropical grasses which belong to the tribes Chlorideae and Paniceae (nine studied) have leaf net photosynthetic rates which are high at normal ambient CO_2 concentration and near-optimal temperatures, and approach light saturation only at illuminances of 10,000 f.c. They have zero CO_2 concentration, no apparent photorespiration, and large parenchyma sheaths surrounding leaf vascular bundles (bundle sheaths). Therefore these grasses can be included in the group which contains some tropical grasses, some species of Cyperaceae, and some dicotyledons, which also have these characteristics as well as the C_4 dicarboxylic acid pathway of CO_2 fixation; [113 species of the Gramineae are now found to have the C_4 pathway].

'Tropical pasture legumes and including Townsville stylo, on the other hand, appear to belong to a group which includes temperate grasses and most dicotyledons. Tropical legumes have no prominent bundle sheaths and have the C_3 Calvin pathway of CO_2 fixation, apparent photorespiration, and a CO_2 compensation concentration of 40 μ^{11-1}.'

The first clue that a mechanism alternative to the Calvin cycle (in which the first radioactive product formed is a 3-carbon acid) may operate in some plants was provided by the observation that 4-carbon acids, malate and aspartate, were the first compounds in which radioactivity was detected (HATCH, 1972). Members of the Gramineae (sugarcane, maize, sorghum and millet and several important grass species) fix CO_2 via the C_4 pathway; subsequently it has been found that the pathway also operates in species from nine other, widely divergent families. Appreciation of the characteristics of the C_4 pathway species which set them apart from Calvin cycle species (Table 12/4) came at a time of increasing interest in tropical species. The operation of the C_4 pathway is associated with an unusual leaf anatomy which is probably essential for the correct functioning of the process. Interrelated with the higher potential for photosynthesis is the greater water-use economy of C_4 species.

BROWN & SMITH (1972) introduce the Kranz syndrome, a group of related anatomical, cytological and physiological characters, including operation of the C_4 pathway, to confirm their theory of the origin in geological time, evolution and present distribution of some members of the Gramineae (9.2.1).

In Java (HOFSTRA, AKSORNKOAE, ATMOWIDJOJO, BANAAG, SANTOSA, SASTROHOETOMO and THU, 1972), plants with a low CO_2 compensation point (C_4 type) appear to be less common than plants with a high CO_2 compensation point (C_3 type). Plants with the C_4 pathway are found only among herbs (50 per cent of herbaceous species found in the open at low altitudes, percentage decreasing with increasing altitude towards zero at 3,000 m.; percentage much lower in shade than in full light).

Table 12/4

Some unusual physiological and performance characteristics specifically associated with the C_4 pathway (HATCH, 1972).

Characteristic	Approximate quantitative relation compared with Calvin cycle species
1. High temperature optimum for photosynthesis	30–45°C versus 15–30°C
2. High light optimum for photosynthesis	Full sunlight versus 30% full sunlight
3. High photosynthesis rates per unit leaf area	Average about twofold under optimal conditions
4. High growth rates under optimal conditions for photosynthesis	Average about twofold under optimal conditions
5. High dry matter production per unit of water used	2–3 fold and more at high temperatures and radiation

All species of Gramineae found in open at low altitudes are of C_4 type; species of the C_3 type are found in shade or at higher altitudes (denser cloud cover?). It is considered that the Paniceae (containing genera of both the C_4 and C_3 types) may be intermediate between the Panicoid/Chloridoid group (C_4) and the Festucoid/Arundoid group (C_3) of BROWN (1958). HOFSTRA's statement that the properties of C_4

plants are assumed to be an adaptation to extreme conditions and that one might expect to find such plants in Java may be queried. It seems unlikely that the present grass flora of Java (MONOD DE FROIDEVILLE, 1968) would find a tropical rain forest ecoclimate 'extreme'.

The ORSTOM Centre d'Adiopodoumé has studied morphological characteristics and CO_2 exchanges in the leaves of *Panicum maximum* (MONTENY, 1971a and b):

The C_4 acids produced at the level of chloroplasts of the chlorenchyma move towards the chloroplasts of the perivascular layer. Here they enter the Calvin cycle, to end up in the form of hexose and starch which can be stored there. It seems that this specification of the functions of the chloroplasts plays a part in the level of photosynthetic efficiency. In the same way, the architecture of the leaf would ensure:

a. rapid elimination of the photosynthesized products towards the storage tissues or growing points, and
b. a better water supply to chlorophyll tissues maintaining chloroplastic activity in spite of certain unfavourable conditions.

In the CSIRO Division of Plant Industry, Canberra, DOWNES (1969, 1970) measured differences in rates of photosynthesis and transpiration in tropical and temperate grasses under controlled conditions. The rates of gas exchange and leaf characteristics have been examined in ten species of *Sorghum*, selected because of diversity of habitat, ploidy, degree of modification by man and taxonomic relations (DOWNES, 1971). *Sorghum arundinaceum* which originated in the low-light environment of equatorial forests was capable of higher rates of photosynthesis and greater stomatal response if it was conditioned to low (pre-treated in half-sunlight) rather than high radiation conditions. In this respect, it differed from the other species, although all have the characteristics of plants with the C_4 pathway of photosynthesis. Generally species from high radiation environments were capable of the highest rate of photosynthesis, although some exceptions were noted. The Australian species occur in savanna woodland and tall grass communities of low latitudes which experience conditions of varying radiation in the rainy season. This appears to be reflected in their relative sensitivity to light-intensity conditioning. Ploidy has little bearing on gas exchange responses. It appears that it is the light regime to which a genotype is adapted that determines the responses detected.

All the species had gas exchange characteristics of species having the C_4 dicarboxylic acid pathway of photosynthesis. A subgenus of *Panicum* differs from the rest of the genus in having features associated with the C_3 (Calvin) pathway (DOWNTON, BERRY & TREGUNNA, 1969); in this subgenus, many species are forest grasses, and specialized plastids are absent from parenchyma bundle sheath cells. There is also a reduction in plastid content of these cells in *S. arundinaceum*. Apparently the presence or absence of the C_4 pathway is not directly associated with evolutionary adaptation to shade, since different pathways are used in forest species of *Panicum* and *Sorghum*.

The classification of *S. arundinaceum* and other species as forest species should be made with care (2.4). Although *S. arundinaceum* may come from a part of Africa where the climax vegetation type is equatorial forest, it is unlikely that a plant of this growth habit belongs in the forest shade; more probably along river banks and on cleared savanna, in full light. If a comparison between forest and non-forest species is to be made, the accepted grasses of forest shade and forest fringes would merit study.

TREHARNE & COOPER (1969) state that it may be that the differing temperature optima for net photosynthesis in temperate and tropical Gramineae are in part associated with the differing temperature sensitivities of the relevant carboxylating enzymes.

DOWNTON & TREGUNNA (1968) determined the carbon dioxide compensation of some Gramineae (including tropical species) and considered these results in relation to the proposed photosynthetic carboxylation pathway (HATCH, SLACK & JOHNSON, 1967, HATCH, 1972), rates of photosynthesis, grass systematics, leaf anatomy and distribution of starch in the leaf. The introduction of temperate and tropical characters into species of the opposite type might extend their range of cultivation. Although genera belonging to the same phylogenetic group have basically the same photosynthetic physiology and internal leaf anatomy, DOWNTON, BERRY & TREGUNNA (1969) were able to find temperate and tropical characteristics in two subgenera of the genus *Panicum*.

12.4 Root studies

'Knowledge of the performance of root systems is both of intrinsic physiological interest and important in agriculture, because the growth of crops is often limited by inadequate water and nutrients or by unfavourable soil conditions' (RUSSELL, 1970). RUSSELL reviews the question under the heads of (a) absorption and translocation of nutrients, (b) micro-organisms, exudates and nutrient uptake, and (c) root systems under field conditions. A simple physical model has been evolved (GREACEN & OH, 1972) which describes a mechanism controlling growth and distribution of roots in the soil. In otherwise uniform soil, roots proliferate in wetter soil zones, because mechanical resistance is low rather than because soil water potential is high.

A study has been made of the root systems of plants in nine vegetation types in the savannas of Surinam described by HEYLIGERS (1963) and VAN DONSELAAR (1965) (see 3.2.2). The growth of roots is determined by the two mutually opposing factors: desiccation and soil saturation (VAN DONSELAAR-TEN BOKKEL-HUININK, 1966). The roots of monocotyledons tolerate the water-saturated condition, but cannot protect themselves against the influences of extended drought. If the water supply comes from beneath, the root systems of monocotyledons are narrow and penetrate deeply; if rain water is the main source, root systems spread. The quicker the rain water penetrates into the soil and falls to lower levels, the larger are the horizontal dimensions of the root systems. Vegetation structure above ground level is related to available rooting volume.

Techniques for studying the roots of tropical grasses and legumes are described by TAERUM & GWYNNE for East Africa (1969), BONZON & PICARD for West Africa (1969), BRAY, HACKER & BYTH for Queensland (1969) and by authors listed in Table 12/5.

12.5 Seasonality in growth

The seasonal availability of feed on tropical grazing lands and seasonal variations in quality represent major problems for the stabilization of animal nutrition in the characteristic climatic sequences of these latitudes.

The wet season development pattern of six perennial and two annual native grasses has been studied in the Northern Territory of Australia (LAZARIDES, NORMAN & PERRY, 1965), numbers of vegetative and reproductive tillers, phases of reproductive development and height being recorded at weekly intervals from early December to late April. The growth patterns of a Tall Grassveld sward in Natal (RETHMAN & BOOYSEN, 1969), measured with an inclined point and by clipping and weighing, were dominated by moisture conditions except in winter when temperature was limiting. Although the inclined point method will not be used in its present form, there is nevertheless a need for a method of yield determination other than clipping and weighing because of the effect of different times of defoliation on the sward.

In the humid tropical environment of the low and middle regions of the Côte d'Ivoire, the response of *Panicum maximum*, *Cynodon plectostachyus*, *Stylosanthes gracilis* and *Centrosema pubescens* to climatic factors is largely determined by (a) the position of the watertable and the state of desiccation in the upper soil layers; (b) the importance and availability of the water reserves, and (c) the type of roots of the plants (TALINEAU, 1970). A neutron moisture gauge has been used in these studies (TALINEAU & LESPINAT, 1971).

Table 12/5

Studies of roots of tropical grasses and legumes

Species	Country	Reference
Brachiaria mutica	Nepal	KHYBRI & MISHRA, 1967
Cenchrus biflorus	India	SHARMA, 1968
Cenchrus ciliaris	Kenya	DAGG, 1969
	Kenya	TAERUM, 1970a and b.
	India	reference under LASIURUS
	India	CHAKRAVARTY & KACKAR, 1967
Cenchrus setigerus	India	reference under LASIURUS
Chloris gayana	Kenya	DAGG, 1969
	Kenya	TAERUM, 1970a and 1970b
Cymbopogon sp.	Nepal	KHYBRI & MISHRA, 1967
Cynodon dactylon	Kenya	DAGG, 1969
Cynodon plectostachyus	Nepal	KHYBRI & MISHRA, 1967
Desmodium uncinatum	Australia	ANDREW, 1966
Dichanthium annulatum	India	reference under LASIURUS
Digitaria decumbens	Venezuela	FRENCH & RODRIGUEZ, 1962
Eragrostis superba	Kenya	TAERUM, 1970a
Glycine javanica	Australia	BRAY, HACKER & BYTH, 1969
Hyparrhenia rufa	Venezuela	FRENCH & RODRIGUEZ, 1962
Imperata cylindrica	Nepal	KHYBRI & MISHRA, 1967
Lasiurus sindicus	India	CHAKRAVARTY & KACKAR, 1967
Lasiurus hirsutus	India	DABADGHAO, MARWAHA, GUPTA, DAS & DEB ROY, 1963
Medicago sativa	Australia	BRAY, HACKER & BYTH, 1969
	Australia	ANDREW, 1966
Melinis minutiflora	Brazil	PAULA, GOMIDE, SYKES & CHAVES,1969
Panicum maximum	Kenya	TAERUM 1970a and 1970b
	Côte d'Ivoire	PICARD, 1969
Panicum antidotale	India	Reference under LASIURUS
Panicum purpurascens	Venezuela	FRENCH & RODRIGUEZ, 1962
Pennisetum purpureum	Venezuela	FRENCH & RODRIGUEZ, 1962
Phaseolus lathyroides	Australia	ANDREW, 1966
Rhynchosia minima	India	CHAKRAVARTY & KACKAR, 1967
Setaria anceps	Australia	BRAY, HACKER & BYTH, 1969
Stylosanthes humilis	Australia	ANDREW, 1966
Themeda triandra	Kenya	DAGG, 1969
	Kenya	TAERUM, 1970a

12.5.1 SEASONALITY IN INDIVIDUAL SPECIES

Variation in chemical composition and nutritive value has been studied: *Melinis minutiflora*, *Digitaria pentzii*, *Panicum maximum* in Venezuela (COELHO DA SILVA & GOMIDE, 1967)
Hyparrhenia rufa in Costa Rica (TERGAS, BLUE & MOORE, 1971)
Hyparrhenia and *Loudetia* spp. in Lamto savanna, Côte d'Ivoire (HÉDIN, 1967)

Andropogon gayanus in Nigeria (HAGGAR, 1970, HAGGAR & AHMED, 1970)
Pennisetum purpureum x P. typhoides (DAFTARDAR & ZENDE, 1968)
Native herbage in Korea (HAN, KIM & PARK, 1970)
Paspalum plicatulum, *Setaria sphacelata* and *Chloris gayana* (HENZELL & OXENHAM, 1964)
Dichanthium annulatum – caloric values and energy structure (GUPTA, 1972)

12.6 **Flowering, reproduction and environment**

EVANS (1969) has reviewed some case histories related to the induction of flowering. TAINTON (1969) finds that tropical/subtropical grasses show a wide range of response to photoperiod. Some of the response patterns are extremely complex and often do not seem to be related to the conditions to which the plants are adapted. Factors of the environment other than photoperiod (possible interactions between daylength, temperature of day and night, and light energy input) may be important in controlling the flowering process.

12.6.1 REPRODUCTION IN INDIVIDUAL SPECIES

Brachiaria mutica	JAVIER, 1970b
Cenchrus ciliaris	EVERS, HOLT & BASHAW, 1969
Chloris gayana	JONES & PRITCHARD, 1971
Dichanthium intermedium	SARAN & DE WET, 1970
Heteropogon contortus	RETHMAN & BOOYSEN, 1968b
	TOTHILL & KNOX, 1968
Panicum maximum	JAVIER, 1970a
Paspalum dilatatum	HICKS & MITCHELL, 1969
Pennisetum clandestinum	MADHAVA RAO & RAMALINGAM, 1964
P. purpureum	JAVIER, 1970b
Setaria sphacelata	BOONMAN, 1972
Themeda australis	EVANS & KNOX, 1969
Th. triandra	TAINTON & BOOYSEN, 1963
Tristachya hispida	TAINTON & BOOYSEN, 1963

12.7 **Seed yield**

Factors affecting seed yield of tropical grasses in Kenya (BOONMAN, 1971a) include:
(i) prolonged head emergence within plants
(ii) prolonged anthesis and stigma exsertion within single heads

(iii) decreasing duration of flowering and decreasing headlength in progressively later emerging heads
(iv) low seed setting per head
(v) low number of head-producing tillers (BOONMAN, 1971b)
(vi) large variation in time of initial head emergence between plants within a cultivar, and
(vii) other indirect limiting factors, including seed shattering, spikelet diseases, bird damage and tendency to lodge.

12.8 Genetical response to physiological stress

Plants become exposed to physiological stress when their habitat changes through natural climatic fluctuations or man-induced desiccation, or when they try to establish themselves in foreign ecological niches with a more severe ecoclimate, or when they are sown in a habitat differing widely from that from which they came.

The first parts of a plant to demonstrate the effects of stress are not the somatic tissues (wilting), but the reproductive organs. They are the most susceptible to the initial loss of turgour following exposure to aridity, or to the reduction in the supply up the vascular system (the nutritional pathway) of the specialized substances required for the development of the gametes. This reaction to stress does not become immediately visible, although fundamental changes are actually occurring at the microscopic level. The timing of the occurrence of water stress is important in relation to yield, nitrogen supply and the variety used, as demonstrated with wheat by D. F. BEECH (CSIRO Division of Land Research Annual Report 1969/70, pp. 42/3 – see Fig. 12/1).

The study by O. H. FRANKEL (CSIRO Division of Plant Industry Annual Report, 1971) and R. B. KNOX (Botany Department, Australian National University) approaches the proposition made below. It has been found that a combination of short day and high temperature subsequent to floral induction intensifies floral defects in basal sterile wheats. Histological and cytochemical studies are designed to pinpoint the developmental condition of individual flowers, and to ascertain the amount and distribution of RNA at flower sites that are known to develop into empty, defective or normal flowers.

In such studies, there is a need to distinguish between the different routes of movement of different classes of nutrients within the plant: (a) water and relatively crude nutrients and salts coming from the root into and up the stem; (b) nutrients from the initial stage of processing in the leaves by the mechanisms associated with the C_3 (Calvin) and C_4 pathways of CO_2 fixation (12.3), passing from the leaves to the reproductive organs; and (c) nutrients processed within the reproductive

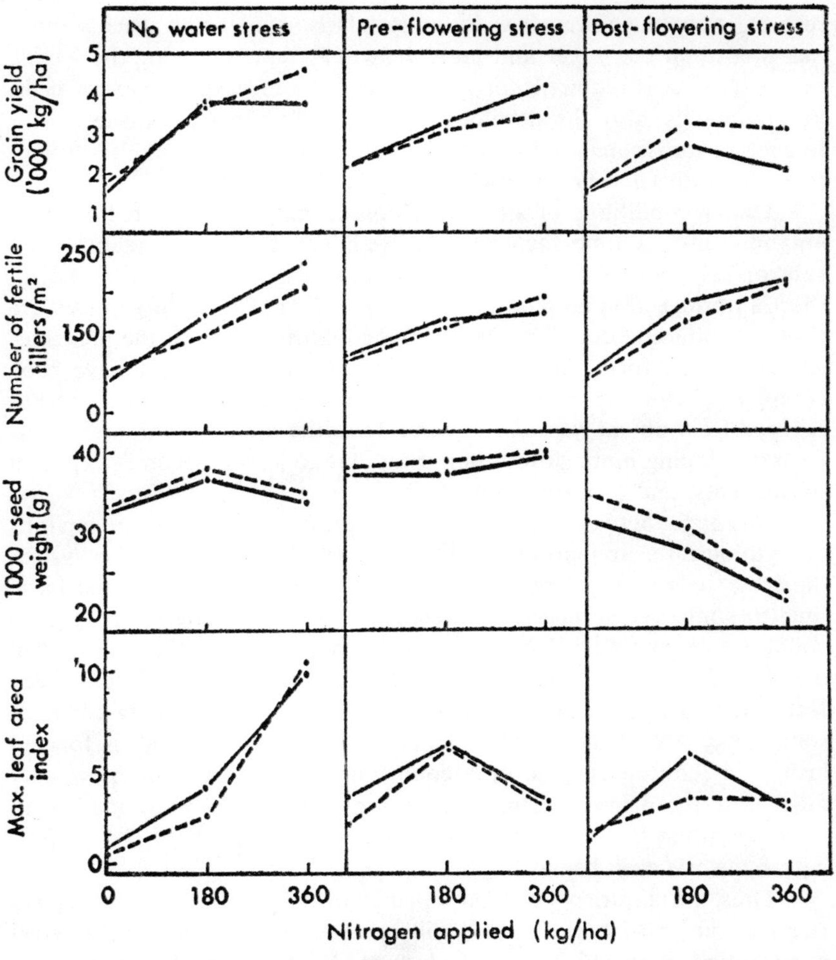

Fig. 12/1: Effect of water stress on Gamenya (———) and Mexico 120 (- - - - - -) wheat. (D. F. BEECH, CSIRO Division of Land Research Annual Report 1969/70).

organs themselves. These last complex components are highly significant for the optimal nutrition and future development of the gametes; they are, however, present in minute amounts for quite short periods, and call for cyto-physiological studies using refined techniques of isolation and analysis.

Physiological stress introduces competition for nutrients between the male and the female organs of bisexual flowers. The exact nature and consequences of that competition are governed by the structure of the flower or inflorescence, and particularly by the morphology of the vascular system. The male organs are always at a disadvantage in this respect, since their branch vascular system is smaller and usually more marginal than the larger and more central system supplying the female organs (which is required not only to feed the megaspore mother cells, but ultimately also the developing zygote). In the Gramineae, a difference in nutritional status is to be expected between spikelets at different positions and heights along the spike.

Under conditions of stress and consequent competition for water and nutrients, a time factor is involved, as observed in plants from subtropical South America growing outdoors against a 'warm' wall in the Cambridge Botanic Garden (WHYTE, 1930). The male gametes are always produced before the female. When stress operates, the processes leading to the formation of mature pollen grains may not have been completed before the female organs begin to draw upon the total but reduced flow up the vascular system. Thus the male system, which has been developing more slowly than usual due to stress, becomes deprived of nutrients, and deterioration sets in.

Although these events may not be visible externally, they are obvious on cytological examination. When the supply from the vascular system fails, the cells of the tapetum, whose task it is to transform the basic nutrients into the special requirements of the developing pollen grains, collapse progressively. This disaster has an immediate effect on pollen development, depending upon the stage of meiosis which has been reached. Some cells may complete meiosis; in others, lagging chromosomes may be left along the spindle when the new cell wall is formed around a reduced chromosome complement; in others, the blow may fall before the paired, partially paired or unpaired chromosomes begin to move along the spindle; the extreme result is death of the nucleus before meiosis even begins.

Thus, the mature pollen from a plant which has suffered physiological stress at this most critical phase will be of a highly variable cytological composition. Some of the changes induced by stress are of such a nature that they will become genetically fixed in an offspring produced after fertilization of a normal female gamete by abnormal pollen.

But again, in case of more severe stress, the processes of meiosis in the ovary may also be affected; all manner of chromosomal irregularities may be bequeathed to the ultimate female gamete, to be fertilized by a pollen grain which is probably itself not normal in its chromosomal complement. One variant of abnormality caused by physiological stress is 'failed meiosis', the characteristic of apomictic plants in which the male organs have failed completely, and the female organs have only partially completed their normal processes, yet sufficiently to provide a basis for a new organism. Irregularities in chromosome pairing at meiosis have been noted in the apomicts, *Urochloa mosambicensis*, *Paspalum plicatulum* and *P. gnenoarum* (PRITCHARD, 1970).

Thus, physiological deficits of water and/or nutrients caused by an adverse P/E ratio or loss of soil fertility become expressed through the structure of a plant in malformation and sterility, first the male, and subsequently the female, reproductive organs. The creation in this way of chromosomally abnormal zygotes is a more acceptable reason for the great intraspecific variation (or variability) of grass species in arid and semi-arid regions, leading to rapid evolution and speciation, than those proposed by STEBBINS (1952) – not that plants evolve specialized structures for adaptation to dry conditions, but that, among the many new forms produced by physiological stress in their ancestors, some will have a new combination of characters favouring their genetic and adaptational stability in the new or changed environment. This also provides a mechanism for increased speciation (when combined with inter-crossing) along the fluctuating borders between ecoclimatic zones, where physiological stress in critical seasons or years is a significant factor.

Addendum

Dr. JORGE H. MORELLO has provided the following particulars regarding text figures 3/2, 3/3, 3/4 and 3/5.

Fig. 3/2, page 44, 3/3, page 45 and 3/5, page 47: Urundaizal - *Astroneum urundeova* is the dominant tree species, with *Tabebuia ipé* and *Schinopsis balansae*

Fig. 3/3, page 45, and 3/5, page 47: Pajonales = tall bunch grass community; Pirizales = heliophytic community with *Cyperus giganteus* and *Scirpus americanus* dominant

Fig. 3/4, page 46: ants = *Atta wollenveideri*

BIBLIOGRAPHY

General

ARBER, A., – 1935 – The Gramineae: A Study of Cereal, Bamboo and Grass. Cambridge at the University Press. 480 pp.

AUBREVILLE, A., – 1962 – Savanisation tropicale et glaciations quaternaires. *Adansonia* 2: 16–84.

AUBREVILLE, A., – 1966a – Les lisières forêt-savane des régions tropicales. *Adansonia* 6: 175–187.

AXELROD, A., – 1970 – Mesozoic palaeogeography and early angiosperm history. *Bot. Rev.* 36: 277–319.

BASHAW, E. C., A. W. HOVIN & E. C. HOLT, – 1970 – Apomixis, its evolutionary significance and utilization in plant breeding. In 'Proceedings of the Eleventh International Grassland Congress,' pp. 245–248.

BEETLE, A. A., – 1957 – Evolution of the grasses. Wyoming Range Man. Iss. no. 110.

BEETLE, A. A., – 1971 – Distribution as a key to the age and origin of grasses. In 'The Natural Classification of the Gramineae,' pp. 111–116, I.U.B.S. Symposium no. 40; in 'Recent Advances in Botany,'. University of Toronto Press.

BEWS, J. W., – 1929 – The World's Grasses: Their Differentiation, Distribution, Economics and Ecology. Longmans, Green and Co., London, New York, Toronto. 408 pp.

BOGDAN, A. V., – 1969 – Rhodes grass: review article. *Herbage Abstracts* 39: 1–13.

BOMMER, D., – 1966 – Futterbau und Grünlandwirtschaft in den Tropen und Subtropen. *Arch. dt. LandwGes.* 37: 152–166.

BOURLIÈRE, F. & M. HADLEY, – 1970 – The ecology of tropical savannas. *Ann. Rev. of Ecol. Syst.* 1: 125–152.

BREESE, E. L., – 1969 – The measurement and significance of genotype-environment interactions in grasses. *Heredity* 24: 27–44.

BROWN, D., – 1954 – Methods of Surveying and Measuring Vegetation. Bulletin no. 42. Commonwealth Agricultural Bureaux. Farnham Royal, 223 pp.

BROWN, W. V., – 1958 – Leaf anatomy in grass systematics. *Bot. Gaz.* 119: 170–178.

BROWN, W. V. & B. N. SMITH, – 1972 – Grass evolution, the Kranz syndrome, $^{13}C/^{12}C$ ratios, and continental drift. *Nature (Lond.)* 239: 345–346.

BURTON, G. W., – 1966 – Breeding better Bermuda grass. In 'Proceedings of the Ninth International Grassland Congress,' pp. 93–96. Departamento da Produção Animal da Secretaria da Agricultura do Estado de São Paulo.

BURTON, G. W., – 1970 – Breeding subtropical species for increased animal production. In 'Proceedings of Eleventh International Grassland Congress,' pp. A56–63. University of Queensland Press, St. Lucia.

BUTTERWORTH, M. H., – 1967 – The digestibility of tropical grasses. *Nutr. Abstr. Rev.* 37: 349–368.

BUTTERWORTH, M. H. & L. J. A. DIAZ, – 1970 – Use of equations to predict the nutritive value of tropical grasses. *J. Range Mgmt.* 23: 55–58.

BUTZIN, F., – 1970a – Die Blattnervatur der Paniceae in ihrer Bedeutung für die systematische Gliederung dieser Tribus. *Willdenowia* 6: 167–178.

BUTZIN, F., – 1970b – Die systematische Gliederung der Paniceae. *Willdenowia* 6: 179–192.

CARLQUIST, S., – 1967 – The biota of long-distance dispersal. V. Dispersal to Pacific islands. *Bull. Torr. Bot. Club* 94: 129–162.

CELARIER, R. P., K. L. MEHRA & M. L. WULF, – 1958 – Cytogeography of the *Dichanthium annulatum* complex. *Brittonia* 10: 59–72.

CHAPMAN, G. P., – 1970 – Patterns of Change in Tropical Plants. University of London Press, London. 112 pp.
CHRISTIAN, C. S., – 1963 – The use and abuse of land and water. In 'The Population Crisis and the Use of World Resources,' Vol. 2, – World Academy of Art and Science, W. Junk, The Hague.
CHRISTIAN, C. S., & G. A. STEWART, – 1968 – Methodology of integrated surveys. In 'Aerial Surveys and Integrated Studies: Proceedings of the UNESCO Conference, Toulouse,' pp. 233–280. UNESCO, Paris.
CLAYTON, W. D., – 1966 – Some aspects of grass taxonomy. In 'Proceedings of the Ninth International Grassland Congress,' pp. 331–333. Departamento da Produção Animal da Secretaria da Agricultura do Estado de São Paulo.
CLIFFORD, H. T., – 1969 – Attribute correlations in the Poaceae (Grasses). *Bot. J. Linn. Soc.* 62: 59–67.
CLIFFORD, H. T., W. T. WILLIAMS & G. N. LANCE, – 1969 – A further numerical contribution to the classification of the Poaceae. *Austral. J. Bot.* 17: 119–131.
COOPER, J. P., – 1970a – Potential production and energy conversion in temperate and tropical grasses (106 references). *Herbage Abstracts* 40: 1–15.
COOPER, J. P., – 1970b – Environmental physiology. In 'Genetic Resources in Plants – their Exploration and Conservation,' ed. O. H. FRANKEL & E. BENNETT, pp. 131–142. Blackwell Scientific Publications, Oxford and Edinburgh.
COOPER, J. P. & N. M. TAINTON, – 1968 – Light and temperature requirements for the growth of tropical and temperate grasses. *Herbage Abstracts* 38: 167–176. (71 references).
CORKILL, L., – 1966 – Breeding for quality in herbage plants. In 'Proceedings of the Ninth International Grassland Congress,' pp. 205–209. Departamento da Produção Animal da Secretaria da Agricultura do Estado de São Paulo.
CVELEV, N. N., – 1969 – Some problems of the evolution of the grasses (Poaceae). *Bot. Z. Leningrad* 54: 361–373.
DANSEREAU, P., – 1951 – Description and recording of vegetation upon a structural basis. *Ecol.* 32: 172–229.
DAVIES, J. G., – 1966 – Pasture improvement in the tropics. In 'Proceedings of Ninth International Grassland Congress,' pp. 217–220. Departamento do Produção Animal da Secretaria da Agricultura do Estado de São Paulo.
DAVIES, J. G. & R. J. JONES, – 1971a – Futterbau-Pflanzen. 1. Dauerweiden. In 'Handbuch der Landwirtschaft und Ernährung in den Entwicklungsländern,' ed. P. VON BLANCKENBURG & H. – D. CREMER, pp. 607–26. Eugen Ulmer, Stuttgart.
DAVIES, J. G. & R. J. JONES, – 1971b – Ackerfutterpflanzen. In 'Handbuch der Landwirtschaft und Ernährung in den Entwicklungsländern,' ed. P. VON BLANCKENBURG & H. – D. CREMER, pp. 627–40. Eugen Ulmer, Stuttgart.
DAVIES, W., – 1960 – Plenary paper. In 'Proceedings of Eighth International Grassland Congress,' pp. 1–7.
DAVIES, W. & C. L. SKIDMORE (eds.), – 1966 – Tropical Pastures. Faber and Faber, London. 215 pp.
DESCOINGS, B., – 1971 – Méthode de description des formations herbeuses intertropicales par la structure de la végétation. *Candollea* 26: 223–257.
DIX, R. L. & R. G. BEIDLEMAN (eds.), – 1969 – The Grassland Ecosystem: a Preliminary Synthesis. Proceedings of the Information Synthesis Project, Grassland Biome, U.S. International Biological Program. Sci. Ser. 2, Range Science Department, Colorado State University, Fort Collins. 441 pp.
DIX, R. L. & R. G. BEIDLEMAN (eds.), – 1970 – The Grassland Ecosystem: a Preliminary Synthesis. A Supplement. United States International Biological Program. Sci. Ser. 2, Suppl. Range Science Department, Colorado State University, Fort Collins. 136 pp.

Downton, W. J. S., J. Berry & E. B. Tregunna, – 1969 – Photosynthesis: temperate and tropical characteristics within a single grass genus. *Science* 163: 78–79.

Downton, W. J. S. & E. B. Tregunna, – 1968 – Carbon dioxide compensation – its relation to photosynthetic carboxylation reactions, systematics of the Gramineae, and leaf anatomy. *Canad. J. Bot.* 46: 207–215 (30 references).

Dyksterhuis, E. J., – 1957 – The savanna concept and its use. *Ecol.* 38: 435–442.

Dyne, G. M. van, – 1970 – A systems approach to grasslands. In 'Proceedings of Eleventh International Grassland Congress,' pp. A131–143. University of Queensland Press, St. Lucia.

Eeckhout, L. E., – 1954 – Contribution à l'uniformisation de la terminologie phytogéographique. In 'Rapp. et Comm. 8me Congrès Int. Bot., Paris, Sect. 7 et 8.' pp. 69–74.

Evans, L. T., – 1969 – The Induction of Flowering: Some Case Histories. Cornell University Press, Ithaca. 488 pp.

Evers, G. W., E. C. Holt & E. C. Bashaw, – 1969 – Seed production characteristics and photoperiodic responses in buffel grass, *Cenchrus ciliaris* L. *Crop. Sci.* 9: 309–310.

FAO – 1971 – Use of Skylab Data for FAO Integrated Resource Surveys. FAO Rome. 46 pp.

Fosberg, F. R., – 1958 – On the possibility of a rational general classification of humid tropical vegetation. In 'Proceedings of UNESCO Symposium on Humid Tropics Vegetation,' pp. 34–59.

Gillard, A., – 1969 – On the terminology of biosphere and ecosphere. *Nature* (*Lond.*) 223: 500–501.

Godron, M. *et al.*, – 1968 – Code pour le Relevé méthodique de la Végétation et du Milieu. C.N.R.S., Paris. 292 pp.

Good, R., – 1964 – The Geography of the Flowering Plants (3rd ed.). Longmans Green and Co. Ltd., London. 518 pp.

Greacen, E. L. & J. S. Oh, – 1972 – Physics of root growth. *Nature* (*Lond.*) 235: 24–25.

Hare, F. K. & J. C. Ritchie, – 1972 – The boreal bioclimates. *Geogr. Rev.* 62: 333–365.

Harlan, J. R., – 1970 – *Cynodon* species and their value for grazing and hay. *Herbage Abst.* 40: 233–238.

Harlan, J. R., – 1971 – Agricultural origins: centers and noncenters. *Science* 174: 468–474.

Harlan, J. R., J. M. J. de Wet & K. M. Rawal, – 1970a – Geographic distribution of the species of *Cynodon* L. C. Rich (Gramineae). *East Afric. Agric. For. J.* 36: 220–226.

Harlan, J. R., J. M. J. de Wet & K. M. Rawal, – 1970b – Origin and distribution of the Seleucidus race of *Cynodon dactylon* (L.) Pers. var. *dactylon* (Gramineae). *Euphytica* 19: 465–469.

Hartley, W., – 1958 – Studies on the origin, evolution and distribution of the Gramineae. I. The tribe Andropogoneae. II. The Paniceae. *Austral. J. Bot.* 6: 116–128, 343–357.

Hartley, W., – 1964 – The distribution of the grasses. In 'Grasses and Grasslands,' ed. C. Barnard, pp. 29–46. Macmillan and Co., London, Melbourne, New York.

Hartley, W. & R. J. Williams, – 1956 – Centres of distribution of cultivated pasture grasses and their significance for plant introduction. Paper no. 17, Seventh International Grassland Congress.

Hatch, M. D., – 1972 – Photosynthesis and the C_4 pathway. In 'CSIRO, Division of Plant Industry Annual Report 1971,' pp. 19–26. Canberra.

Hatch, M. D., C. R. Slack & H. S. Johnson, – 1967 – Further studies on a new pathway of photosynthetic carbon dioxide fixation in sugar-cane and its occurrence in other plant species. *Biochem. J.* 102: 417–422.

Hicks, D. H. & K. J. Mitchell, – 1969 – Flowering in pasture grasses. 2. *Paspalum dilatatum* Por. *New Zealand J. Bot.* 7: 167–172.

Hills, T. L. & R. E. Randall (eds.), – 1968 – The Ecology of the Forest/Savanna Boundary. Proceedings of I.G.U. Humid Tropics Symposium, 1964. Savanna Research Series. McGill University, 128 pp.

Holdridge, L. R., – 1947 – Determination of world plant formations from simple climatic data. *Science* 105: 367–368.

Hubbard, C. E., – 1968 – Grasses (2nd ed.) Penguin Books Ltd., Harmondsworth. 463 pp.

Humphrey, R. R., – 1966 – Range condition analysis – a practical ecological approach to the classification of rangelands. In 'Proceedings of Ninth International Grassland Congress,' pp. 1437–1442. Departamento da Produção Animal da Secretaria da Agricultura do Estado de São Paulo.

International Union of Biological Sciences – 1971 – The natural classification of the Gramineae. Symposium no. 40. In 'Recent Advances in Botany,' pp. 91–145. University of Toronto Press.

Ivens, G. W., – 1972 – The problem of bush in rangeland. *Span* 15: 23–26.

Jaeger, F., – 1945 – Zur Gliederung und Benennung des tropischen Graslandgürtels. *Verh. Naturforsch. Ges. Basel* 56: 509–520.

Kawanabe, S. & T. Hirakawa, – 1971 – Comparative ecology of warm-type and cool-type grasses. *Japanese J. Anim. Husb.* 25: 1181–1186, 1315–1318; 1442–1444; 1570–1574 (in Japanese).

Lauer, W., – 1952 – Humide und aride Jahreszeiten in Afrika und Südamerika und ihre Beziehung zu den Vegetationsgürteln. *Bonner Geogr. Abhandl.* 9: 15–98.

Legris, P., – 1972 – A new formula of hydric balance. *Tropical Ecology* 13: 12–26.

Lieth, H., – 1972 – Modelling the primary productivity of the world. (i) *Nature and Resources* 8(2): 5–10. (ii) *Tropical Ecology* 13: 125–130. (iii) Über die Primärproduktion der Pflanzendecke der Erde. *Angew. Bot.* 1.

Long, G., – 1966 – Modern phytosociological and ecological methods and their use in range research inventory. In 'Proceedings of Ninth International Grassland Congress,' pp. 1349–1352. Departamento da Produção Animal da Secretaria da Agricultura do Estado de São Paulo.

Long, G., – 1969 – Perspectives nouvelles de la cartographie biogéographique végétale intégrée. *Vegetatio* 18: 44–63.

Löve, D., – 1963 – Dispersal and survival of plants. In 'North Atlantic Biota and their History,' ed. Löve A. & D. Löve, pp. 189–205. Pergamon Press, Oxford, London, New York, Paris. 430 pp.

Lugo, A. E., – 1969 – Energy flow in some tropical ecosystems. In *Proc. Soil Crop Soc. Fla* 29: 254–64.

MacFadyen, A., – 1964 – Energy flow in ecosystems and its exploitation by grazing. In 'Grazing in Terrestrial and Marine Environments. A Symposium of the British Ecological Society, 1962,' ed. D. J. Crisp, pp. 3–20. Blackwell, Oxford.

McKell, C., J. P. Blaisdell & J. R. Goodin (technical eds.), – 1972 – Wildland Shrubs – Their Biology and Utilization. Intermountain Forest and Range Experiment Station, USDA Forest Service General Technical Report INT-1, Ogden, Utah. 494 pp.

McMillan, C., – 1969 – Ecotypes and ecosystem function. *BioScience* 19: 131–134.

Mears, P. T., – 1970 – Kikuyu (*Pennisetum clandestinum*) as a pasture grass – a review. *Trop. Grasslands* 4: 139–52.

Melville, R., – 1966 – Continental drift. Mesozoic continents and the migrations of the angiosperms. *Nature* (*Lond.*) 211: 116–120.

Melville, R., – 1972 – Personal communication.

Morley, F. H. W. & C. R. W. Spedding – 1968 – Agricultural systems and grazing experiments. *Herbage Abstracts* 38: 279–287.

MOTT, G. O., B. R. EDDLEMAN & D. H. TIMM., 1970 – Systems approach to a study of the soil-plant-animal complex in the tropics. *Proc. Soil Crop Soc. Fla* 29: 238–253.
NESTEL, B. L. & M. J. CREEK, – 1962 – Pangola grass – a review. *Herbage Abstracts* 32: 265–271.
NORRIS, D. O., – 1966a – Rhizobium relationships in legumes. In 'Proceedings of Ninth International Grassland Congress,' pp. 1087–1092. Departamento da Produção Animal da Secretaria da Agricultura do Estado de São Paulo.
NORRIS, D. O., – 1966b – The legumes and their associated Rhizobium. In 'Tropical Pastures,' ed. W. DAVIES & C. L. SKIDMORE, pp. 89–105. Faber and Faber, London.
NORRIS, D. O., – 1967 – The intelligent use of inoculants and lime pelleting for tropical legumes. *Trop. Grasslands* 1: 107–121.
NORRIS, D. O., – 1970 – Nodulation of pasture legumes. In 'Australian Grasslands', ed. R. M. MOORE, pp. 339–348. Australian National University Press, Canberra.
NORRIS, D. O., – 1971 – Leguminous plants in tropical pastures. Paper to Seminar on Forage Crop Research in West Africa, University of Ibadan. 34 pp.
Papua and New Guinea: Administration of Territory of/UNESCO, Southeast Asia, – 1960 – Symposium on the Impact of Man on Humid Tropics Vegetation. 402 pp.
PERRY, R. A., – 1964 – Value of ecological surveys. In 'Proceedings and Papers of the IUCN Ninth Technical Meeting, 1963,' Part IV, 'Ecological Research and Development,' pp. 303–312.
PERRY, R. A., – 1967 – Integrated Surveys of Pastoral Areas. S20 ITC/UNESCO, Delft. 27 pp.
POLHILL, R. M., – 1972 – Personal communication.
PRAT, H., – 1960 – Vers une classification naturelle des Graminées. *Bull. Soc. bot. France* 107: 32–79.
PRESCOTT, J. A., J. A. COLLINS & G. R. SHIRPURKAR, – 1952 – The comparative climatology of Australia and Argentina. *Geogr. Rev.* 42: 118–133.
PROCTOR, V. W., – 1968 – Long-distance dispersal of seeds by retention in digestive tract of birds. *Science* 160: 321–322.
RICHARDS, P. W., – 1952 – The Tropical Rain Forest: An Ecological Study. Cambridge University Press, London. 450 pp.
RICHARDS, P. W., – 1961 – The types of vegetation of the humid tropics in relation to the soil. In 'Tropical Soils and Vegetation: Proceedings of the Abidjan Symposium,' pp. 15–23. UNESCO, Paris.
RIJKS, D. A. & J. T. WALKER, – 1968 – Evaluation and computation of potential evaporation in the tropics. *Exper. Agric.* 4: 351–357.
RINEY, T. A., – 1970 – Wildlife and nomadic stocks in semi-arid lands. In 'World Review of Animal Production,' pp. 27–37. FAO, Rome.
ROSS, H. H., – 1972 – The origin of species diversity in ecological communities. *Taxon* 21: 253–259.
RUSSELL, R. S., – 1970 – Root systems and plant nutrition – some new approaches. *Endeavour* 29 (107): 60–66.
SARAN, S. & J. M. J. DE WET, – 1970 – The mode of reproduction in *Dichanthium intermedium* (Gramineae). *Bull. Torrey bot. Club* 97: 6–13.
SCHIMPER, A. F. W., – 1903–1904 – Plant Geography upon a Physiological Basis. Oxford.
SCHNELL, R., – 1970 – 1971 – Introduction à la phytogéographie des Pays tropicaux. Vol. 1. Les Flores, les Structures. 500 pp. Vol. 2. Les Milieux, les Groupements végétaux. pp. 501–952. Gauthier-Villars, Paris.
SINGH, D. N., – 1966 – Classification of the Gramineae in relation to cytology. *Advg. Front. Pl. Sci.* 17: 197–303.
SODERSTROM, T. R. & C. E. CALDERÓN, – 1971 – Insect pollination in tropical rain forest grasses. *Biotropica* 43: 890–905.

Spedding, C. R. W., – 1971 – Grassland Ecology. Clarendon Press, Oxford, Oxford University Press, London. 221 pp.

Stebbins, G. L., – 1952 – Aridity as a stimulus to plant evolution. *Amer. Naturalist* 86: 33–44.

Stebbins, G. L., – 1956 – Cytogenetics and evolution of the grass family. *Am. J. Bot.* 43: 890–905.

Stebbins, G. L., – 1959 – A suggested classification of grass genera of North America based on modern criteria. In 'Second Int. Bot. Congress' vol. II, p. 378.

Stewart, G. A., – 1970 – High potential productivity of the tropics for cereal crops, grass forage crops and beef. *J. Austral. Inst. Agric. Sci.* 36: 85–101.

Stewart, G. & S. S. Hutchings, – 1936 – The point-observation-plot (sq.-foot density) method of vegetation survey. *J. Amer. Soc. Agron.* 28: 714–726.

Summerhayes, V. S., – 1931 – An enumeration of the angiosperms of the Seychelles Archipelago. *Trans. Linn. Soc. Ser. Biol.* 19: 261–299.

Takhtajan, A., – 1969 – Flowering Plants, Origin and Dispersal. Translated C. Jeffrey. Oliver and Boyd, Edinburgh. 310 pp.

Tateoka, T., – 1957a – Miscellaneous papers on the phylogeny of Poaceae. *J. Jap. Bot.* 32: 275–287.

Tateoka, T., – 1957b – Karyotaxonomic studies in Poaceae. Annual Report of National Institute of Genetics, Japan 8: 37–38.

Tateoka, T., – 1960 – Cytology in grass systematics: a critical review. *Nucleus* 3: 81–110.

Taubert, P., – 1894 – Leguminosae. In 'Die Natürlichen Pflanzenfamilien,' ed. Engler & Prantl, vol. 3, pp. 70–388. Wilhelm Englemann, Leipzig.

Thomas, P. T., – 1966 – Herbage breeding at the crossroads. In 'Proceedings of the Ninth International Grassland Congress,' pp. 211–214. Departamento da Produção Animal da Secretaria da Agricultura do Estado de São Paulo.

Tothill, J. C., – 1966 – Phenological variation in *Heteropogon contortus* and its relation to climate. *Austral. J. Bot.* 14: 35–47.

Tothill, J. C., – 1968 – Variation and apomixis in *Heteropogon contortus*, Gramineae. *Bol. Soc. Argentina Bot.* 12: 188–201.

Tothill, J. C., – 1970 – Reproductive behaviour as it affects the distribution and patterns of response of *Heteropogon contortus* in tropical and subtropical regions. In 'Proceedings of the Eleventh International Grassland Congress,' pp. 30–32. University of Queensland, St. Lucia.

Tothill, J. C. & R. B. Knox, – 1968 – Reproduction in *Heteropogon contortus*. 1. Photoperiodic effects on flowering and sex expression. *Austral. J. Agric. Res.* 19: 869–878.

Treharne, K. L. & J. P. Cooper, – 1969 – Effect of temperature on the activity of carboxylases in tropical and temperate Gramineae. *J. Exp. Bot.* 20: 170–175.

Troll, C., – 1952 – Das Pflanzenkleid der Tropen in seiner Abhängigkeit von Klima, Boden und Mensch. Tagungsber. wiss. Abhandl. Deutscher Geogr. Tag. Frankfurt, 1951. pp. 35–66. Remagen.

Tuley, P., – 1968 – *Stylosanthes gracilis*: a review. *Herbage Abstracts* 38: 87–94.

Tutin, T. G., – 1958 – Classification of the legumes. In 'Nutrition of the Legumes,' ed. E. G. Hallsworth, pp. 3–14. Butterworth, London.

UNESCO, – 1971 – International grazing lands programme. *Nature and Resources* 7: 21.

Valentine, D. H., – 1970 – Evolution at zones of vegetational transition. *Feddes Repert.* 81 (1–5): 33–39.

Veldkamp, J. F., – 1971 – Personal communication.

Verdcourt, B., – 1972 – Personal communication.

Walter, H., – 1964 – Die Vegetation der Erde, Band 1. Die Tropischen und Subtropischen Zonen. Gustav Fisher Verlag, Jena. Translated D. MUELLER DOMBOIS, ed. J. H. BURNETT, – 1971 – Ecology of Tropical and Subtropical Vegetation. Oliver and Boyd, Edinburgh. 539 pp.

WALTER, H., – 1969 – El problema de la sabana: investigaciones ecofisiológicas en el Africa Sur-Occidental en comparación con las condiciones existentes en Venezuela. *Boln Soc. Venez. cienc. nat.* 28 (115–116): 123–144.

WALTER, H., – 1970 – Ergänzende Betrachtungen zu der im Klimadiagramm-Weltatlas Verwendeten Klimadarstellung. *Erdkunde* 24: 145–149.

WALTER, H., – 1971a – Biosphäre, Produktion der Pflanzendecke und Stoffkreislauf in ökologisch-geographischer Sicht. *Geographische Zeitschrift* 59: 116–130.

WALTER, H., – 1971b – Die natürlichen Standortverhältnisse und ihre Verbesserung. 1. Klima und Vegetation. In 'Handbuch der Landwirtschaft und Ernährung in den Entwicklungsländern,' ed. P. VON BLANCKENBURG & H. D. CREMER, vol. 2, pp. 1–25. Verlag Eugen Ulmer, Stuttgart.

WEBB, L. J., – 1966 – An ecological comparison of forest-fringe grassland habitats in eastern Australia and eastern Brazil. In 'Proceedings of Ninth International Grassland Congress,' pp. 321–330. Departamento da Produção Animal da Secretaria da Agricultura do Estado de São Paulo.

WEISCHET, W., – 1969 – Klimatologische Regeln zur Vertikalverteilung der Niederschläge in Tropengebirgen. *Die Erde* 100: 287–306.

WHITTAKER, R. H., – 1972 – Evolution and measurement of species diversity. *Taxon* 21: 213–251.

WHYTE, R. O., – 1930 – Sterility and floral abnormality in the tetraploid *Saxifraga potternensis*. *J. Genetics* 23: 93–121.

WHYTE, R. O., – 1968a – Grasslands of the Monsoon. Faber and Faber, London/ F. A. Praeger, New York. 325 pp.

WILLIAMS, O. B., – 1968 – That uneasy state between animal and plant in the manipulated situation. *Proc. Ecol. Soc. Austral.* 3: 167–174.

WILLIS, J. C., – 1922 – Age and Area: A Study in Geographical Distribution and Origin of Species. Cambridge University Press, Cambridge. 259 pp.

WILLIS, J. C. & J. S. GARDINER, – 1931 – Flora of the Chagos Archipelago. *Trans. Linn. Soc. Ser. Biol.* 19: 301–306.

WOODFORD, E. K., – 1971 – Presidential address. *J. Brit. Grassland Soc.*: 26: 1–7.

ZONNEVELD, I. S., – 1972a – Land evaluation and land(scape) science. In 'ITC Textbook of Photo-Interpretation 7. Use of Aerial Photographs in Geography and Geomorphology,' chapter 7: 4, 106 pp. ITC, Enschede.

ZONNEVELD, I. S., – 1972b – Personal communication.

Africa South of the Sahara

ACOCKS, J. P. H., – 1953 – Veld types of South Africa. Botanical Survey of South Africa Memoir no. 28. 191 pp.

ADEGBOLA, A. A., B. O. ONAYINKA & J. K. EWEJE, – 1968 – The management and improvement of natural grassland in Nigeria. *Niger. Agric. J.* 5: 5–6.

ADEJUWON, J. O., – 1970 – The ecological status of coastal savannahs in Nigeria. *J. Trop. Geogr.* 30: 1–10.

ADEJUWON, J. O., – 1971 – A biogeographical survey of the dynamics of savanna vegetation in Nigeria. *Nigerian Geogr. J.* 14: 31–48.

ANDREAE, B., – 1966 – Weidewirtschaft im südlichen Afrika. Erdkundliches Wissen, Heft 15. Franz Steiner Verlag, Wiesbaden. 49 pp.

ANDREAE, B., – 1971 – Das Instrumentarium des Futterausgleiches in extensiven Weidewirtschaften und Dornbuschsteppen und Trockensavannen. *Ber. Landw.* 49: 84–112.

ASARE, E. O., – 1970 – Effects of fertilization, height and frequency of cutting on herbage yield and nutritive value of *Cenchrus ciliaris* Linn. (buffel grass) in the forest region of Ghana. In 'Proceedings of Eleventh International Grassland Congress', pp. 594–597. University of Queensland, St. Lucia.

ASTLE, W. L., – 1966 – The edaphic grasslands of Zambia. In 'Proceedings of Ninth International Grassland Congress,' pp. 363–373. Departamento da Produção Animal da Secretaria da Agricultura do Estado de São Paulo.

ASTLE, W. L., – 1968–69 – The vegetation and soils of Chishinga Ranch, Luapula Province, Zambia. *Kirkia* 7: 73–102.

AUBREVILLE, A., – 1968 – Les étranges mosaïques forêt-savane du sommet de la boucle de l'Ogooué au Gabon. *Adansonia* 7: 13–22.

BADER, F., – 1967a – Probleme der Erfassung und Darstellung der Vegetation an ausgewählten Teilgebieten des Kartenblattes Lake Victoria. *Die Erde* 98: 142–149.

BADER, F., – 1967b – Die Vegetationsgeographie auf dem Blatt Lake Victoria in Afrika-Kartenwerk der Deutschen Forschungsgemeinschaft. *Berichten der Deutschen Botanischen Gesellschaft* 80: 291–297.

BADER, F., – 1974 – Die Vegetation Ostafrikas. Deutsches Forschungsgemeinschaft.

BARRAULT, J., – 1971 – Work to date in Northern Cameroon on *Pennisetum pedicellatum.* Paper to Seminar on Agricultural Research in West Africa on Forage Crops. University of Ibadan. 20 pp.

BARRAULT, J., A. V. BOGDAN, M. BORGET *et al.*, – 1971 – A summary of discussions at seminar on forage crop research in West Africa, University of Ibadan, 26–30 April, 1971. 18 pp.

BARTTHA, R., – 1970 – Plantes fourragères de la Zone sahélienne d'Afrique. Weltforum Verlag, Munich. 306 pp.

BAWDEN, M. G., – 1965 – A reconnaissance of the land resources of eastern Bechuanaland. *J. appl. Ecol.* 2: 357–365.

BAWDEN, M. G. & D. M. CARROLL, – 1968 – The Land Resources of Lesotho. Land Resource Study no. 3. Land Resources Division, Directorate of Overseas Surveys, Tolworth, Surrey. 89 pp.

BAWDEN, M. G., D. M. CARROLL & P. TULEY, – 1972 – The Land Resources of North East Nigeria. 3. The Land Systems. Land Resource Study no. 9. Overseas Development Administration, Foreign and Commonwealth Office, Surbiton, Surrey. 466 pp.

BELL, R. H. V., – 1971 – A grazing ecosystem in the Serengeti. *Scientific American* 225: 86–93.

BILLE, J. C., – 1965 – Evolution des pâturages naturels des hauts plateaux de la

République Centrafricaine en exploitation traditionnelle Bororo. *Rev. d'Elev. Méd. Vétér. Pays tropicaux.* 18: 313–316.

BILLE, J. C., G. LAMARQUE, J.-P. LEBRUN & R. RIVIÈRE, – 1968 – Etude agrostologique des pâturages de la région des savanes (République du Togo). Etude Agrostologique no. 24. Institut d'Elevage et de Médecine Vétérinaire des Pays Tropicaux, Maisons Alfort. 108 pp.

BILLE, J. C., J.-P. LEBRUN & R. RIVIÈRE, – 1968/69 – Pâturages et Cultures fourragères en Afrique intertropicale. Institut d'Elevage et de Médicine Vétérinaire des Pays Tropicaux, Maisons Alfort/Secretariat d'Etat aux Affaires Etrangères chargé de la Coopération, Paris. 218 pp.

BOGDAN, A. V., – 1959 – The selection of tropical ley grasses in Kenya: general considerations and methods. *East Afric. Agric. For. J.* 24: 206–217.

BOGDAN, A. V., – 1963 – Three interesting introductions from the local grass flora at Kitale, Kenya. *J. Brit. Grassland Soc.* 18: 247–48.

BONZON, B. & D. PICARD – 1969 – Matériel et méthodes pour l'étude de la croissance et du développement en pleine terre des systèmes racinaires. *Cah. ORSTOM, ser. biol.* 9: 3–18 (24 references).

BOONMAN, J. G., – 1971a – Experimental studies on seed production of tropical grasses in Kenya. 1. General introduction and analysis of problems. *Neth. J. agric. Sci.* 19: 23–36.

BOONMAN, J. G., – 1971b – 2. Tillering and heading in seed crops of eight grasses. *Neth. J. agric. Sci.* 19: 237–249.

BOONMAN, J. G., – 1972 – 3. The effect of nitrogen and row width on seed crops of *Setaria sphacelata* cv Nandi II. *Neth. J. agric. Sci.* 20: 22–34.

BOOYSEN, P. DE V., N. M. TAINTON & J. D. SCOTT – 1963 – Shoot-apex development in grasses and its importance in grassland management. *Herbage Abstracts* 33: 209–213. (25 references).

BORGET, M., – 1971 – IRAT forage crop research in equatorial Africa and Madagascar. Paper to Seminar on Agricultural Research in West Africa, University of Ibadan. Institut de Recherches agronomiques tropicales et des Cultures Vivrières. 25 pp.

BOUDET, G., – 1966 – Investigation and cartographic method for tropical pastures applied at the Breeding and Veterinary Medicine Institute of Tropical Countries. In 'Proceedings of Ninth International Grassland Congress', pp. 475–477. Departamento da Produção Animal da Secretaria da Agricultura do Estado de São Paulo.

BOUDET, G., – 1970 – Management of savannah woodland range in West Africa. In 'Proceedings of Eleventh International Grassland Congress,' pp. 1–3. University of Queensland Press, St. Lucia.

BOUDET, G. & F. BAEYENS, – 1963 – Une méthode d'étude et de cartographie des pâturages tropicaux. *Rev. d'Elev. Méd. vét. Pays Trop.* 16: 191–219.

BOUDET, G. & J. F. ELLENBERGER, – 1971 – Etude agrostologique du Berceau de la Race N'Dama dans le Cercle de Yanfolila (République du Mali). Etude agrostologique no. 30. I.E.M.V.T., Maisons-Alfort. 175 pp.

BOUDET, G. & R. RIVIÈRE, – 1968 – Practical use of fodder analysis to assess the value of tropical pastures. *Rev. d'Elev. Méd. vét. Pays trop.* 21: 227–66.

BOUDET, G., R. RIVIÈRE, J. CLÉMENSAT, J. PAGOT & J. F. LAHORE, – 1961 – Les possibilités fourragères de *Digitaria umfolozi* en zone soudanienne. *Rev. d'Elev. Méd. vét. Pays trop.* 14: 449–68.

BOURREIL, P., – 1967 – Sur l'écologie, la germination et la culture de quelques graminées africaines du genre *Aristida* L. *Annls. Fac. Sci. Marseille* 39: 45–87.

BOURREIL, P., – 1968 – Sur un indice d'appréciation de l'appétence des herbivores pour certaines plantes de la Province du Darfour (Soudan). *J. Agric. trop. Bot. appl.* 15: 200–201.

BOURREIL, P. & H. GILLET, – 1971 – Synthèse des connaissances et des recherches nouvelles sur *Aristida rhiniochloa*, graminée africaine amphitropicale. *Mitt. Bot. Staatssamml., Munich* 10: 309–340.

BOX, T. W., – 1968 – Range resources of Somalia. *J. Range Mgmt.* 21: 388–392.

BRENAN, J. P. M., – 1965 – The geographical relationships of the genera of Leguminosae in tropical Africa. *Webbia* 19: 545–578.

BROWN, L. H. & J. COCHEMÉ, – 1969 – A Study of the Agroclimatology of the Highlands of Eastern Africa. FAO/UNESCO/WMO Interagency Agroclimatology Project. FAO, Rome. 330 pp.

CASEBEER, R. L., – 1969 – Integrated resource surveys for rangeland development in Kenya. FAO Ad Hoc Technical Conference on Grassland Production and Fodder Management in Africa south of Sahara, Nairobi, 1969. Paper PL: GFA/69/2. FAO, Rome. 5 pp.

CASEBEER, R. L., – 1971 – Wildlife and Land Use in Kenya. Report prepared for the Government of Kenya by FAO. AGP: SF/KEN 11. Technical Report no. 3. FAO, Rome. 22 pp.

CASEBEER, R. L., & G. G. KOSS, – 1970 – Food habits of wildebeest, zebra, hartebeest and cattle in Kenya Masailand. *East Afric. Wildlife J.* 8: 25–36.

CHHEDDA, H. R., – 1971 – *Cynodon* improvement in Nigeria. Paper to Seminar on Forage Crop Research in West Africa, University of Ibadan. 12 pp.

CLARK, J. D., – 1970 – The Prehistory of Africa. Thames and Hudson, London. 302 pp.

COCHEMÉ, J. & P. FRANQUIN, – 1967 – An Agroclimatology Survey of a Semiarid Area in Africa south of the Sahara. WMO Technical Note no. 86. WMO, Geneva. 136 pp.

COETZEE, W. H. K., P. L. LE ROUX, D. S. DU TOIT & P. J. J. VAN RENSBURG – 1970 – Relationship between fodder and ruminal composition and its possible application to the nutritive evaluation of natural pastures V. Ruminal composition as a possible criterion of the nutritive value of natural pastures. *Agroanimalia* 2: 151–156.

COMBES, D., & J. PERNES, – 1970 – Variations dans les nombres chromosomiques du *Panicum maximum* Jacq. en relation avec le mode de reproduction. *C.r. hébd. Séanc. Acad. Sci., Paris (D)* 270: 782–785.

COMPÈRE, R., – 1968 – Exploitation rationnelle de la prairie permanente d'altitude à *Exotheca abyssinica* Anders. et *Eragrostis boehmii* Hack. au Rwanda et au Burundi. *Bull. Rech. agron. Gembloux* 3: 583–604.

CONSEIL SCIENTIFIQUE POUR L'AFRIQUE AU SUD DU SAHARA, – 1956 – Rapport final de la Réunion de Spécialistes en Phytogéographie/Final Report of the Specialist Meeting on Phytogeography, Yangambi, Belgian Congo. Joint Secretariat, Bukavu. 379+32 pp.

DAGG, M., – 1969 – Hydrological implications of grass root studies at a site in East Africa. *J. Hydrology* 9: 438–444.

DAGG, M. & J. R. BLACKIE, – 1970 – Estimates of evaporation in east Africa in relation to climatological classification. *Geogr. J.* 136: 227–234.

DAGG, M., T. WOODHEAD & D. A. RIJKS, – 1970 – Evaporation in east Africa. *Bull. Intern. Assoc. Sci. Hydrology* 15: 61–67.

DOUGALL, H. W. & A. V. BOGDAN, – 1958 – The chemical composition of the grasses of Kenya. *East Afric. agric. For. J.* 24: 17–23.

DUVIGNEAUD, P., – 1951 – Les savanes du Bas-Congo: essai de phytosociologie topographique. *Lejeunia, Rev.-Bot. Liège* 10: 1–192.

DUVIGNEAUD, P., – 1958 – La végétation du Katanga et de ses sols métallifères. *Bull. Soc. Roy. Bot. Belg.* 90: 127–286.

DUVIGNEAUD, P. & S. DENAEYER DE SMET, – 1963 – Cuivre et végétation au Katanga. *Bull. Soc. Roy. Bot. Belg.* 96: 93–231.

EAST AFRICAN ROYAL COMMISSION – 1955 – 1953–1955 Report. H.M.S.O., London. 482 pp.

ELDIN, M., – 1969 – Utilisation de la dimérisation de l'Anthracène sous l'action du rayonnement dans le Proche Ultraviolet pour une mesure de l'énergie lumineuse. Application à des mesures sous couverts végétaux. *J. West Afric. Sci. Assoc.* 14: 17–25.

ELDIN, M., – 1970 – Détermination de l'évapotranspiration d'un couvert végétal à partir de profils hydriques établis à la sonde à neutrons. In 'Peaceful Uses of Atomic Energy in Africa,' pp. 133–139. International Atomic Energy Agency, Vienna.

FERNANDES, A. B., – 1971 – Interspecific hybridisation of *Pennisetum purpureum* Schumach. with *Pennisetum typhoides* Stapf and Hubb. Paper to Agricultural Research Seminar on Forage Crops, University of Ibadan, 9 pp.

FIELD, C. R., – 1970 – Observations on the food habits of tame warthog and antelope in Uganda. *East Afr. Wildl. J.* 8: 1–17.

FOOD AND AGRICULTURE ORGANIZATION OF THE UNITED NATIONS, – 1970 – Improvement of Livestock and Dairy Industry, Malawi. Report for Government of Malawi, based on work of H. H. HEADY. AGA: SF/MLW 5, Technical Report 3. FAO, Rome. 15 pp.

FAO/UNDP, – 1971 – Wildlife and Land Use in Kenya. Range Management Division of Ministry of Agriculture, Kenya, based on work of R. L. CASEBEER. AGP: SF/KEN 11. Technical Report no. 3. FAO, Rome. 22 pp.

GAFF, D. F., – 1971 – Desiccation-tolerant flowering plants in southern Africa. *Science* 174: 1033–1034.

GARNIER, P., J. LAHORE & P. DUBOIS, – 1968 – Etude du pâturage naturel à Madagascar: productivité, conséquences pratiques. *Rev. Elev. Méd. vét. Pays trop.* 21: 203–218.

GLEDHILL, D., – 1966 – Carpet grasses (*Axonopus* spp.) in West Africa. In 'Proceedings of Ninth International Grassland Congress,' pp. 469–470. Departamento da Produção Animal da Secretaria da Agricultura do Estado de São Paulo.

GLOVER, P. E., – 1947 – A provisional Checklist of British and Italian Somaliland Trees, Shrubs and Herbs. Crown Agents, for the Government of Somaliland, London. 446 pp.

GRANIER, P., – 1965 – Le rôle de l'élevage extensif dans la modification de la végétation à Madagascar. *Rev. Elev. Méd. vét. Pays trop.* 18: 293–305.

GRANIER, P. & J. LAHORE, – 1966 – Amélioration des pâturages. Le *Brachiaria brizantha*. *Rev. Elev. Méd. vét. Pays trop.* 19: 233–242.

HAGGAR, R. J., – 1970 – Seasonal production of *Andropogon gayanus*. 1. Seasonal changes in yield components and chemical composition. *J. agric. Sci. Camb.* 74: 487–494.

HAGGAR, R. J. & M. B. AHMED, – 1970 – Seasonal production of *Andropogon gayanus*. 2. Seasonal changes in digestibility and feed intake. *J. Agric. Sci. Camb.* 75: 369–373.

HARRINGTON, G. N., – 1973 – Fire ecology of the savanna grasslands of Ankole, Uganda. *J. appl. Ecol.* (in press).

HEDBERG, O., – 1971a – Evolution of the Afroalpine flora. In 'Adaptive Aspects of Insular Evolution,' ed. W. L. STERN, pp. 16–23. Washington State University Press.

HEDBERG, O., – 1971b – The high mountain flora of the Galama Mountain in Arussi Province, Ethiopia. *Webbia* 26: 101–128.

HÉDIN, L., – 1967 – Recherches écologiques dans la savane de Lamto (Côte d'Ivoire). La valeur fourragère de la savane. *La Terre et la Vie*: 21: 249–61.

HOPKINS, B., – 1968 – Vegetation of the Olokemeji Forest Reserve, Nigeria. 5. The vegetation on the savanna site with special reference to its seasonal changes. *J. Ecol.* 56: 97–115.

HOPKINS, B., – 1970a – Vegetation of the Olokemeji Forest Reserve, Nigeria. 6. The

plants on the forest site with special reference to their seasonal growth. *J. Ecol.* 58: 765–793.
HOPKINS, B., – 1970b – Vegetation of the Olokemeji Forest Reserve, Nigeria. 7. The plants on the savanna site with special reference to their seasonal growth. *J. Ecol.* 58: 795–825.
I'ONS, J. H., – 1967 – Veld improvement in Swaziland through the introduction of a subtropical legume. *Proc. Grassland Soc. South Africa* 2: 71–73.
I'ONS, J. H., – 1968 – The development of tropical pastures for the Swaziland middleveld. *Proc. Grassland Soc. South Africa* 3: 67–73.
I'ONS, J. H., – 1969 – The effect of introducing stylo (*Stylosanthes guyanensis*) into veld on animal production in the middleveld of Swaziland. *Proc. Grassland Soc. South Africa* 4: 113–115.
JACQUES-FÉLIX, H., – 1962 – Les Graminées d'Afrique tropicale. I. Généralités, Classification, Description des Genres. Bull. Sci. no. 8, IRAT, Paris. 345 pp.
JACQUES-FÉLIX, H., – 1968 – Evolution de la végétation au Cameroun sous l'influence de l'homme. *J. Agric. trop. Bot. appl.* 15 (7–8): 350–356.
KANNEGIETER, A., – 1966 – The cultivation of grasses and legumes in the forest zone of Ghana. *Ghana J. Sci.* 6 (3–4): 97–109.
KAYSER, K., W. MANSHARD, H. MENSCHING & J. H. SCHULTZE, – 1966 – Das Afrika-Kartenwerk. Ein Schwerpunkt-Program der Deutschen Forschungsgemeinschaft. *Die Erde* 97: 85–95.
KEAY, R. W. J., – 1959 – Derived savanna: derived from what? *Bull. d'IFAN Ser. A.* 21: 427–438.
KENWORTHY, J. M., – 1966 – Temperature conditions in the tropical highland climates of East Africa. *East Afr. Geogr. Rev.* 4: 1–11.
LAMOTTE, M., – 1967 – Recherches écologiques dans la savane de Lamto (Côte d'Ivoire): Présentation du milieu et du programme de travail. *La Terre et la Vie* 21: 197–215.
LANGDALE-BROWN, I., H. A. OSMASTON & J. G. WILSON, – 1964 – The Vegetation of Uganda and its Bearing on Land Use. Government Printer, Entebbe. 147 pp.
LANSBURY, T. J., R. ROSE-INNES & G. L. MABEY, – 1965 – Studies on Ghana grasslands: yield and composition on the Accra Plains. *Trop. Agri. Trin.* 42: 1–18.
LEMON, P. C., – 1968a – Effects of fire on an African plateau grassland. *Ecol.* 49: 316–322.
LEMON, P. C., – 1968b – Fire and wildlife grazing on an African plateau. In 'Proceedings of Annual Tall Timbers Fire Ecology Conference no. 8', pp. 71–88.
LETOUZEY, R., – 1968 – Etude phytogéographique du Cameroun. Ed. P. Lechevalier, Paris. 511 pp.
LONG, M. I. E., W. K. NDYANABO, B. MARSHALL & D. D. THORNTON, – 1969 – Nutritive value of grasses in Ankole and the Queen Elizabeth National Park, Uganda. 4. Mineral content. *Trop. Agri. Trin.* 46: 201–209.
LONG, M. I. E., D. D. THORNTON & B. MARSHALL, – 1969 – Nutritive value of grasses in Ankole and the Queen Elizabeth National Park, Uganda. 2. Crude protein, crude fibre and soil nitrogen. *Trop. Agri. Trin.* 46: 31–42.
LONG, M. I. E., D. D. THORNTON, W. K. NDYANABO, B. MARSHALL & H. SSEKAB, – 1970 – The mineral status of dairy farms in the parts of Buganda and Busoga bordering Lake Victoria, Uganda. 2. Nitrogen and mineral content of pastures. *Trop. Agri. Trin.* 47: 37–49.
MCKAY, A. D., – 1970 – Range management research at EAAFRO. *E. Afr. agric. For. J.* 35: 346–349.
MCKAY, A. D. & P. E. FRANDSEN, – 1969 – Chemical and floristic components of the diet of Zebu cattle (*Bos indicus*) in browse and grass range pastures in semi-arid upland area of Kenya. 1. Crude protein. *Trop. Agri. Trin.*: 46: 279–292.

MOLL, E. J., – 1968 – A plant ecological reconnaissance of the Upper Mgeni Catchment. *J. South Afric. Bot.* 34: 401–420.

MONTENY, B. A., – 1971a – Caractéristiques morphologiques et Echanges en Gaz carbonique des Feuilles de *Panicum maximum*. ORSTOM, Adiopodoumé. 28 pp.

MONTENY, B., – 1971b – Mesure de la Productivité d'un Couvert végétal par une Méthode gravimétrique. ORSTOM, Adiopodoumé. 27 pp.

MONTENY, B. A., D. COMBES & M.-F. PRÉVOST, – 1971 – Etude morphologique du Limbe foliaire de *Panicum maximum* en Relation avec la Photosynthèse. ORSTOM, Adiopodoumé. 19 pp. (23 references).

MORRISON, J., – 1966 – Productivity of grass and grass/legume swards in the Kenya Highlands. *East Afric. Agric. For. J.* 32: 19–24.

MORRISON, J., – 1969 – Grasses and legumes at high altitudes in tropical Africa. *Herbage Abstracts* 39: 101–106.

MWAKHA, E., – 1970 – Observations on the growth of Bungoma grass (*Entolasia imbricata* Stapf) *CSIRO Plant Introduction Rev.* 7: 24–29.

OJO, O., – 1969 – Potential evapotranspiration and the water balance in West Africa. An alternative method of Penman. *Arch. Met. Geophys. Bioklim. Ser. B.* 17: 239–260.

OKIGBO, B. N. & H. R. CHHEDDA, – 1966 – Natural fertility and chromosome numbers in several strains of star grasses. *Niger. Agric. J.* 3: 72–75.

OLIVER, J., – 1969 – Problems of determining evapotranspiration in the semi-arid tropics illustrated with reference to the Sudan. *J. trop. Geogr.* 28: 64–74.

OWEN, M. A. & H. W. BRZOSTOWSKI, – 1966 – A grass cover for the upland soils of the Kongwa Plain, Tanganyika. *Trop. Agric. Trin.* 43: 303–314.

PAGOT, J. R., – 1971 – Pâturages naturels, Cultures fourragères et Elevage dans les Régions tropicales de l'Afrique francophone. Institut d'Elevage et de Médecine vétérinaire des Pays tropicaux, Maisons Alfort. 11 pp.

PERNES, J. & D. COMBES, – 1970 – Incidence des systèmes de multiplication sur la répartition et la variabilité phénotypique du *Panicum maximum* Jacq. en Côte d'Ivoire. *Cah. ORSTOM, sér. biol.* 14: 13–34.

PERNES, J., D. COMBES & R. RÉNÉ-CHAUME, – 1970 – Phytogénétique – Différenciation des populations naturelles du *Panicum maximum* Jacq. en Côte d'Ivoire par acquisition de modifications transmissibles, les unes par graines apomictiques, d'autres par multiplication végétative. *C.R. Acad. Sc. Paris* 270: 1992–1995.

PETERSON, R. A., – 1969 – Analysis of pastoral problems. Paper to Ad Hoc Technical Conference on Grassland Production and Fodder Management in Africa south of the Sahara. PL: GFA/69/5, FAO, Rome. 19 pp.

PEYRE DE FABRÈGUES, B., – 1965 – Etudes et principes d'exploitation de pâturage de steppe en République du Niger. *Rev. Elev. Méd. vét. Pays trop.* 18: 329–332.

PICARD, D., – 1969 – Comparaison de deux techniques de prélèvement d'échantillons de racines. *Cah. ORSTOM, sér. biol.* 9: 19–31.

PICARD, D. &. C. FILLONNEAU, – 1971 – Production de quelques plantes fourragères en Côte d'Ivoire, en fonction de différents modes d'exploitation. Communication au Colloque sur l'Intensification des Productions fourragères en Zone tropicale humide et l'Utilisation par les Ruminants. Guadeloupe. 12 pp.

PIOT, J., – 1966 – Etudes pastorales en Adamaoua Camerounais. *Rev. Elev. Méd. vét. Pays trop.* 19: 45–62.

PIOT, J., – 1969 – Végétaux ligneux et pâturages des savanes de l'Adamaoua au Cameroun. *Rev. Elev. Méd. vét. Pays trop.* 22: 541–559.

PIOT, J., – 1970 – Pâturage aérien au Cameroun. Utilisation des ligneux par les bovins. *Rev. Elev. Méd. vét. Pays trop.* 23: 503–517.

PRATT, D. J., – 1968 – Rangeland development in Kenya. *Ann. arid Zone* 7: 177–208.

Pratt, D. J., – 1969 – Management of arid rangeland in Kenya. *J. Brit. Grassland Soc.* 24: 151–157.

Pratt, D. J., – 1971 – Bush-control studies in the drier areas of Kenya. 6. Effects of fenuron (3-phenyl-1, 1-dimethylurea). *J. appl. Ecol.* 8: 239–245.

Pratt, D. J., P. J. Greenway & M. D. Gwynne, – 1966 – A classification of East African rangeland. *J. appl. Ecol.* 3: 369–382.

Pratt, D. J., & J. Knight, – 1971 – Bush-control studies in the drier areas of Kenya. 5. Effects of controlled burning and grazing management on *Tarchonanthus/Acacia* thicket. *J. appl. Ecol.* 8: 217–237.

Rains, A. B., – 1970 – The evaluation of African rangeland using aerial photographs. In 'Proceedings Eleventh International Grassland Congress,' pp. 99–101. University of Queensland Press, St. Lucia.

Rains, A. B. & A. M. Yalala, – 1972 – The Central and Southern State Lands, Botswana. Land Resource Study no. 11. Land Resources Division, Tolworth Tower, Surbiton, Surrey. 118 pp.

Ramsay, J. M. & R. Rose-Innes, – 1963 – Some quantitative observations on the effects of fire on the Guinea savanna vegetation of Northern Ghana over a period of eleven years. *African Soils* 8: 41–85.

Rattray, J. M., – 1960 – The Grass Cover of Africa. FAO Agricultural Studies no. 49. FAO, Rome. 168 pp. with map.

Rattray, J. M., – 1962 – Criteria used for the measurement of condition and trend in grazing resources. In 'Interdepartmental Conference of Pasture Workers,' pp. 44–67. Government of Southern Rhodesia, Bulawayo.

Rethman, N. F. G. & P. de V. Booysen, – 1968a – The influence of time of defoliation on the vigour of a tall grassveld sward in the next season. *Proc. Grassland Soc. South Africa* 3: 91–94.

Rethman, N. F. G., & P. de V. Booysen, – 1968b – Growth and development in *Heteropogon contortus* tillers. *South Africa J. agric. Sci.* 11: 259–271.

Rethman, N. F. G. & P. de V. Booysen, – 1969 – The seasonal growth patterns of a tall grassveld sward. *Proc. Grassland Soc. South Africa* 4: 56–60.

Risopoulos, S. A., – 1966 – Management and use of grasslands. Democratic Republic of the Congo. Pasture and Fodder Crop Studies no. 1. FAO, Rome. 150 pp.

Roberts, B. R., – 1970 – Assessment of veld condition and trend. *Proc. Grassland Soc. South Africa* 5: 137–139.

Ross, I. C. & G. N. Harrington, – 1969 – The practical aspects of implementing a controlled burning scheme in the Kedepo Valley National Park (second year of operation). *East Afr. Wildl. J.* 7: 39–42.

Sands, E. B., D. B. Thomas, J. Knight & D. J. Pratt, – 1970 – Preliminary selection of pasture plants for the semi-arid areas of Kenya. *East Afr. agric. For. J.* 36: 49–57.

Schalkwyk, A. van, P. E. Lombard & L. F. Vorster, – 1968 – Evaluation of the nutritive value of a *Themeda triandra* pasture in the central Orange Free State. 1. Botanical composition and nutritive value. 2. Comparison between quadrat and simulated grazing samples. 3. Grazing samples and selecting ability of cattle. 4. Digestibility. *South Afric. J. Agri. Sci.* 11: 113–122; 249–258; 483–492; 679–686.

Scott, J. D., – 1970 – Pros and cons of eliminating veld burning. *Proc. Grassland Soc. South Africa* 5: 23–26.

Sillans, R., – 1958 – Les savanes de l'Afrique central française. *Encyclopédie Biol.* 55: 423 pp. Paris.

Southern Rhodesia, Government of, – 1962 – Proceedings of the First Interdepartmental Conference of Pasture Workers. Bulawayo. 118 pp.

Staples, R. R., – 1930 – Studies in veld management. A second report on certain veld burning and grazing experiments. Sci. Bull. 91, Union South Africa Dep. Agric.

STEINKE, T. D. & P. DE V. BOOYSEN, – 1968 – The regrowth and utilisation of carbohydrate reserves of *Eragrostis curvula* after different frequencies of defoliation. *Proc. Grassland Soc. South Africa* 3: 105–110.

STOBBS, T. H., 1969a – The effect of grazing resting land upon subsequent arable crop yields. *East Afr. agric. For. J.* 35: 28–32.

STOBBS, T. H., – 1969b – Animal production from *Hyparrhenia* grassland oversown with *Stylosanthes gracilis* (*S. guyanensis*). *East Afr. agric. For. J.*: 35: 128–134.

STOBBS, T. H., – 1969c – The value of *Centrosema pubescens* (Benth.) for increasing animal production and improving soil fertility in northern Uganda. *East Afr. agric. For. J.* 35: 197–202.

STOBBS, T. H., – 1969d – Beef production from pasture leys in Uganda. *J. Brit. Grassland Soc.* 24: 81–86.

STOBBS, T. H., – 1969e – The effect of grazing management upon pasture productivity in Uganda. 4. Selective grazing. *Trop. Agri. Trin.* 46: 303–309.

STOBBS, T. H., – 1970 – The use of liveweight-gain trials for pasture evaluation in the tropics (final paper in series in same Journal from vol. 21 onwards). 6. A fixed stocking rate design. *J. Brit. Grassland Soc.* 25: 73–77.

STRICKLAND, R. W., – 1972 – Plant collecting mission to Africa 1971–72. *CSIRO Plant Introduction Review* 9: 17–21.

TAERUM, R., – 1970a – Comparative shoot and root growth studies of six grasses in Kenya. *East Afric. agric. For. J.* 36: 94–113. (25 references).

TAERUM, R., – 1970b – A study of root and shoot growth on three grass species in Kenya. *East Afric. agric. For. J.*: 36: 155–170. (21 references).

TAERUM, R., – 1970c – A note on chemical content of some East African grasses. *East Afric. agric. For. J.* 36: 171–176.

TAERUM, R. & M. D. GWYNNE, – 1969 – Methods for studying grass roots under East African conditions and some preliminary results. *East Afric. agric. For. J.*: 35: 55–65.

TAINTON, N. M., – 1969 – Environmental control of flowering in tropical-subtropical grasses. *Proc. Grassland Soc. South Africa* 4: 49–55.

TAINTON, N. M., – 1971 – An analysis of the objectives of resting grassveld. *Proc. Grassland Soc. South Africa* 6: 50–54.

TAINTON, N. M. & P. DE V. BOOYSEN, – 1963 – The effects of management on apical bud development and seeding in *Themeda triandra* and *Tristachya hispida*. *South Afric. J. Agric. Sci.* 6: 21–30.

TAINTON, N. M., P. DE V. BOOYSEN & J. D. SCOTT, – 1970 – Response of tall grassveld to different intensities, seasons and frequencies of clipping. *Proc. Grassland Soc. South Africa* 5: 32–41.

TALBOT, LEE M., H. P. LEDGER & W. J. A. PAYNE, – 1961 – The possibility of using wild animals for animal production in the semi-arid tropics of East Africa. In 'Festschrift zum VIII Intern. Tierzuchtkongress in Hamburg, 1961,' pp. 205–210. Verlag Eugen Ulmer, Stuttgart.

TALINEAU, J. C., – 1970 – Action des facteurs climatiques sur la production fourragère en Côte d'Ivoire. *Cah. ORSTOM. sér. biol.* 14: 51–76 (28 references).

TALINEAU, J. C. & P. A. LESPINAT, – 1971 – Evolution des profils hydriques rélevés par la méthode neutronique sous quelques plantes fourragères en saison sèche. *Cah. ORSTOM, Ser. biol.* 15: 1–19. (20 references).

THOMAS, A. S., – 1966 – The importance of African aquatic grasses. In 'Proceedings of Ninth International Grassland Congress,' pp. 1397–1399. Departamento da Produção Animal da Secretaria da Agricultura do Estado de São Paulo.

THOMAS, D. B. & D. J. PRATT, – 1967 – Bush-control studies in the drier areas of Kenya. 4. Effects of controlled burning on secondary thicket in upland *Acacia* woodland. *J. appl. Ecol.* 4: 325–335.

THORNTON, D. D., M. I. E. LONG & B. MARSHALL, – 1968 – Nutritive value of grasses in Ankole and the Queen Elizabeth National Park, Uganda. 1. Collection sites, soils and grasses. *Trop. Agric. Trin.* 45: 257–267.

THORNTON, D. D., M. I. E. LONG & B. MARSHALL, – 1969 – The mineral status of dairy farms in the parts of Buganda and Busoga bordering Lake Victoria, Uganda. 1. General ecology and soils. *Trop. Agric. Trin.* 46: 269–277.

THORNTON, D. D. & B. MARSHALL, – 1971 – The mineral status of dairy farms in east Uganda. 1. Ecology and soils. *Trop. Agric. Trin.* 48: 217–224.

TRAPNELL, C. G. & J. N. CLOTHIER, – 1957 – The soils, vegetation and agricultural systems of north-western Rhodesia. Government Printer, Lusaka. 87 pp.

TRAPNELL, C. G. & J. F. GRIFFITHS, – 1960 – The rainfall-altitude relation and its ecological significance in Kenya. *East Afric. agric. For. J.* 25: 207–213.

TRAPNELL, C. G., J. D. MARTIN & W. ALLAN, – 1950 – Vegetation-Soil Map of Northern Rhodesia, with accompanying Memorandum. Government Printer, Lusaka.

TROCHAIN, J. L., – 1957 – Accord interafricain sur la définition des types de végétation de l'Afrique tropicale. *Bull. Inst. Et. Centraf.* (13–14): 55–93.

UNESCO, – 1961 – Tropical Soils and Vegetation. Proceedings of the Abidjan Symposium, 1959. UNESCO/CCTA. UNESCO, Paris. 115 pp.

VALENZA, J., – 1970 – Survey of different types of natural pasture land in the Senegal Republic. In 'Proceedings Eleventh International Grassland Congress,' pp. 78–82. University of Queensland Press, St. Lucia.

VERBOOM, W. C., – 1966 – The grassland communities of Barotseland. *Trop. Agric. Trin.* 43: 107–115.

VERBOOM, W. C. & M. A. BRUNT, – 1970 – An Ecological Survey of Western Province, Zambia, with Special Reference to the Fodder Resources. Volume 2. The Grasslands and their Development. Land Resource Study no. 8. Land Resources Division, Directorate of Overseas Surveys, Tolworth, Surrey. 133 pp.

VESEY-FITZGERALD, D. F., – 1970 – The origin and distribution of valley grasslands in East Africa. *J. Ecol.* 58: 51–76.

VINCENT, V., – 1962 – Means and criteria for establishing vegetation type, condition and trend. In 'Interdepartmental Conference of Pasture Workers,' pp. 23–43. Government of Southern Rhodesia, Bulawayo.

VORSTER, L. F. & J. W. C. MOSTERT, – 1968 – Veld fertilization trends over a decade in the central Orange Free State. *Proc. Grassland Soc. South Africa* 3: 111–119.

WILD, H., – 1965 – The flora of the Great Dyke of Southern Rhodesia with special reference to the serpentine soils. *Kirkia* 5: 49–86.

WILD, H., – 1968–69a – Geobotanical anomalies in Rhodesia. 1. The vegetation of copper bearing soils. *Kirkia* 7: 1–71.

WILD, H., – 1968–69b – Geobotanical anomalies in Rhodesia. 2. A geobotanical anomaly occurring in graphitic soils. *Port. Acta biol.* (*b*) 9: 291–299.

WILD, H., – 1970 – Geobotanical anomalies in Rhodesia. 3. The vegetation of nickel-bearing soils. *Kirkia* 7: suppl. 1–62.

WOODHEAD, T., – 1970 – Confidence limits for seasonal rainfall, their value in Kenya agriculture. *Expl. Agric.* 6: 81–86.

WOODHEAD, T. & E. S. WAWERU and E. F. LAWES, – 1970 – Expected rainfall and Kenya agriculture – confidence limits for large areas at minimum cost. *Expl. Agric.* 6: 87–97.

ZAMBIA, REPUBLIC OF – 1970 – Annual Report of the Research Branch. Ministry of Rural Development. 140 pp.

ZONNEVELD, I. S., P. N. DE LEEUW & W. G. SOMBROEK, – 1971 – An ecological interpretation of aerial photographs in a savanna region in northern Nigeria. ITC, Enschede. Series B, no. 63. 41 pp.

America, South, Central and Caribbean

ABIUSSO, N. G., – 1969 – Composición quimica y valor alimenticio de *Salix humboldtiana* Willd. y *Salix babylonica* L. *Revista de la Facultad de Agronomía, La Plata* 44: 159–167.

ABIUSSO, N. G., – 1970 – Analysis quimicos y valor forrajero de plantas indigenas y cultivadas en la Republica Argentina. *Revista de la Facultad de Agronomía, La Plata* 46: 1–14.

ABIUSSO, N. G. & A. G. SCHULZ – 1970 – Composición quimica de plantas indigenas y cultivadas del sector este de la provincia del Chaco. *Revista de Investigaciones Agropecuarias Ser.* 2, *Biología y Producción Vegetal* 7: 1–66.

ADENIYI, S. A. & P. N. WILSON, – 1960 – Studies on Pangola Grass at I.C.T.A., Trinidad. 1. Effects of fertilizer application at time of establishment, and cutting interval on the yield of ungrazed Pangola grass. *Trop. Agric. Trin.* 37: 271–282.

ANDERSON, R. C., – 1969 – Environmental scaling of microhabitats for some Costa Rican grasses. *Turrialba* 19: 57–65.

ATKINSON, W. T., – 1970 – High altitude-low latitude forage plants from Mexico and Latin America. In 'Proceedings of Eleventh International Grassland Congress,' pp. 181–184. University of Queensland Press, St. Lucia.

AUBREVILLE, A., – 1965 – Les étranges savanes des Llanos de l'Orénoque. *Adansonia* 5: 3–13.

AUBREVILLE, A., – 1966b – Le Costa Rica; quelques aspects du pays, de son climat, de sa végétation et de sa flore. *Adansonia* 7: 29–54.

BAZAN, R., – 1972 – Informe de Instituciones Internationales que participen en el desarrollo y fomento de Pastos y Forrajes tropicales: Instituto Interamericano de Ciencias Agricolas, Turrialba. FAO/ICEA-CPAT mimeo. 12 pp.

BEARD, J. S., – 1944 – Climax vegetation in tropical America. *Ecology* 25: 127–158.

BEARD, J. S., – 1953 – The savanna vegetation of northern tropical America. *Ecol. Monogr.* 23: 149–215.

BLYDENSTEIN, J., – 1961 – La vegetación de la Estación Biologica de los Llanos. *Bol. Soc. Venez. Cienc. Natur.* 22: 208–212.

BLYDENSTEIN, J., – 1962 – La sabana de *Trachypogon* del Alto Llano. *Bol. Soc. Venez. Cienc. Natur.* 23: 139–206.

BLYDENSTEIN, J., – 1963 – Cambios en la vegetación despues de protección contra el fuego. 1. El aumento anual en material vegetal en varios sitios quemados y no quemados en la Estación Biologica. *Bol. Soc. Venez. Cienc. Natur.* 23: 233–244.

BLYDENSTEIN, J., – 1967 – Tropical savanna vegetation of the Llanos of Colombia. *Ecology* 48: 1–15.

BLYDENSTEIN, J., – 1968 – Burning and tropical American savannas. In 'Proceedings Annual Tall Timbers Fire Ecology Conference no. 8,' pp. 1–14.

BLYDENSTEIN, J., – 1971 – Recursos Forrajeros de Venezuela. AGP: PFC/19. FAO, Rome. 18 pp.

BLYDENSTEIN, J., – 1972 – Evaluación de recursos naturales. FAO/ICA–CPAT/6. 6 pp.

BRAGADIN, E. A., – 1959 – Las pasturas en la Región de los Llanos. *Revista Agronomica Noroeste Argentino* 3: 289–333.

BULLER, R. E., S. ARONOVICH, L. R. QUINN & W. V. A. BISSCHOFF, – 1970 – Performance of tropical legumes in the upland savannah of Central Brazil. In 'Proceedings Eleventh International Grassland Congress,' pp. 143–146. University of Queensland, St. Lucia.

CAIN, S. A. & G. M. DE O. CASTRO, – 1959 – The Manual of Vegetation Analysis. Harper and Brothers, New York. 337 pp.

CHICCO, C. F., C. RODRIGUEZ & C. E. FUENMAYOR, – 1971 – Efecto de la fertilización

con nitrogeno sobre el rendimento, consumo y digestibilidad del heno de Pangola (*Digitaria decumbens* Stent). *Agronomía Tropical* 21: 215–227.
Coelho da Silva, J. F. & J. A. Gomide, – 1967 – Efeito do estádio de naturação sôbre o consumo e digestibilidade aparente da matéria sêca de três gramíneas tropicais. *Revista Ceres* 13: (76): 255–275.
Cole, M. M., – 1960 – Cerrado, caatinga and pantanal: the distribution and origin of the savanna vegetation of Brazil. *Geogr. J.* 126: 168–179.
Contreras Arias, A., – 1942 – Mapa de las provincias climatologicas de la República Mexicana. Sec. Agricola y Fomento, Dir. de Geografía, Meteorología y Hidrología, Inst. Geografico, México.
Crowder, L. V., – 1971 – Forage and Pasture Research in Tropical America. Paper to Seminar on Forage Crop Research in West Africa, University of Ibadan. 23 pp.
Cuany, R. L., – 1967 – Variability for growth type and flowering characters in 'jaragua', *Hyparrhenia rufa* (Nees.) Stapf. *Agron. Abstr.* Madison p. 8.
Daubenmire, R., – 1972a – Ecology of *Hyparrhenia rufa* (Nees.) in derived savanna in north-western Costa Rica. *J. appl. Ecol.* 9: 11–23.
Daubenmire, R., – 1972b – Some ecologic consequences of converting forest to savanna in northwestern Costa Rica. *Tropical Ecology* 13: 31–51.
Degras, L., – 1968 – Introduction and study of *Digitaria* varieties in the West Indies. *Annls. Amél. Pl.* 18: 159–169.
Donselaar, J. van, – 1965 – An ecological and phytogeographic study of northern Surinam savannas. *Wentia* 14: 1–163.
Donselaar, J. van, – 1968 – Phytogeographic notes on the savanna flora of Southern Surinam (South America). *Acta Bot. Neerl.* 17: 393–404.
Donselaar, J. van, – 1969 – Observations on savanna vegetation-types in the Guianas. *Vegetatio* 17: 271–312.
Donselaar-Ten Bokkel Huinink, W. A. E. van, – 1966 – Structure, root systems and periodicity of savanna plants and vegetations in Northern Surinam. *Wentia* 17: 1–162.
Eden, M. J., – 1970 – Savanna vegetation in the northern Rupununi, Guyana. *J. Trop. Geogr.* 30: 17–28.
Eidt, R. C., – 1968 – The climatology of South America. In 'Biogeography and Ecology in South America,' vol. 1, ed. E. J. Fittkau, J. Illies, H. Klinge, G. H. Schwabe & H. Sioli, pp. 54–81. W. Junk, The Hague.
Eiten, G., – 1972 – The cerrado vegetation of Brazil. *Bot. Rev.* 38: 201–341. (14 pp. bibliography).
Ellenberg, H. von, – 1962 – Wald in der Pampa Argentiniens? *Veröff. d. Geobotanischen Inst. d.E.T.H., Stiftung Rübel in Zürich* (Fest. F. Firbas), 37: 39–56.
Fanshawe, D. B., – 1952 – The Vegetation of British Guyana: A Preliminary Review. Institute Paper no. 29. Imperial Forestry Institute, Oxford. 96 pp.
Fernandes A. P. M., J. A. Gomide & J. M. Braga, – 1970 – Efeito da adubação potássica sôbre a produção e valor nutritivo de algumas gramineas forrageiras tropicais. *Experientiae* 10: 185–208.
FAO, – 1961 – Report of First Meeting of FAO Working Party on Pasture and Fodder Development in Tropical America. FAO, Rome, 46 pp.
FAO/UNSF, – 1966 – Reconocimiento edafologico de los Llanos Orientales: Colombia. Tomo 6. La vegetación natural y la ganaderia en los Llanos Orientales. Sección primera. La vegetación. FAO/SF: 11/COL. FAO, Rome. 233 pp.
French, M. H. & C. Rodriguez, – 1962 – Desarrollo comparativo de las raices de algunos pastos en diferentes regiones de Venezuela. *Agronomia Tropical* 12: 113–122.
Fretes, R., R. Samudio, & C. Gay, – 1970 – Las praderas naturales del Paraguay. 1. Clasificación y descripción. Programa Nacional de Investigación y Extensión Ganadera. Asunción. 86 pp.

Gomide, J. A. & A. C. Tardin, – 1969 – Valor nutritivo do capim-Guatemala (*Tripsacum* sp.) em diferentes idades. *Revista Ceres* 16: 141–147.

Goodland, R. J. A. – 1966 – On the savanna vegetation of Calabozo, Venezuela, and Rupununi, British Guiana. *Boln. Soc. Venez. Cienc. Nat.* 26: 341–359.

Goodland, R. J. A., – 1970 – Plants of the cerrado vegetation of Brazil. *Phytologia* 20: 57–78.

Goodland, R. J. A., – 1971a – A physiognomic analysis of the 'cerrado' vegetation of central Brasil. *J. Ecol.* 59: 411–419.

Goodland, R. J. A., – 1971b – The cerrado oxisols of the Triângulo Mineiro, Central Brazil. *An. Acad. brasil. Cienc.*, 43: 407–414.

Goodland, R. & R. Pollard – 1973 – The Brazilian cerrado vegetation: a fertility gradient. *J. Ecol.* 61: 219–224.

Grof, B., – 1972 – Centro Internacional de Agricultura Tropical: Informes de Instituciones Internacionales que participan en el desarrollo y fomento de Pastos y Forrajes Tropicales. FAO/ICA- CPAT/2 5 pp.

Grossman, J., S. Aronovich & E. do C. B. Campello, – 1966 – Grasslands of Brazil. In 'Proceedings of the Ninth International Grassland Congress,' pp. 39–80. Departamento da Produção Animal da Secretaria da Agricultura do Estado de São Paulo.

Hernandez Efrain, M. F., – 1963 – Los tipos de vegetación de México y su clasificación. Instituto de Biología de la U.N.A.M. y al Escuela Nacional de Agricultura, Chapingo. 29–195.

Heyligers, P. C., – 1963 – Vegetation and soil of a white-sand savanna in Suriname. Mededelingen van het Botanisch Museum en Herbarium van Rijksuniversiteit te Utrecht, no. 191. 148 pp.

Hodges, E. M., G. B. Gillinger, J. E. McCaleb, O. C. Ruelke *et al.*, – 1967 – Pangolagrass. Bulletin 718, Agricultural Experiment Stations, University of Florida, Gainesville. 31 pp.

Horrell, C. R., – 1972 – Informes regionales sobre problemas, actividades y programas crecientes de desarrollo en el campo de pastos y forrajes tropicales: Región Centroamericana. FAO/CIAT–CPAT/10. 15 pp.

Howard, R. A., – 1971 – The 'alpine' plants of the Antilles. In 'Adaptive Aspects of Insular Evolution,' ed. W. L. Stern, pp. 24–28. Washington State University Press.

Hueck, K., 1953 – Urlandschaft, Raublandschaft und Kulturlandschaft in der Provinz Tucumán in nordwestlichen Argentine. *Bonner Geogr. Abh.* 10: 102 pp.

Hymowitz, T., – 1971 – Collection and evaluation of tropical and subtropical Brazilian forage legumes. *Trop. Agric. Trin.* 48: 308–315.

Lanjouw, J., – 1936 – Studies on the vegetation of the Suriname savannahs and swamps. *Ned. Kruidk. Arch.* 46: 823–851.

León, J., – 1971 – Germplasm collections in Central America. *Plant Genetic Resources Newsletter, FAO*, no. 26: 10–19.

Leopold, A. S., – 1950 – Vegetation zones of Mexico. *Ecology* 31: 501–18.

Lima, D. de A., – 1966 – Vegetation of Brazil. In 'Proceedings of Ninth International Grassland Congress,' pp. 29–38. Departamento da Produção Animal da Secretaria da Agricultura do Estado de São Paulo.

Loxton, R. F., G. K. Rutherford & J. Spector, – 1958 – Soil and Land-Use Surveys, no. 2. British Guiana. The Rupununi Savannas. Regional Research Centre of the British Caribbean at the Imperial College of Tropical Agriculture, Trinidad. 33 pp.

McNaughton, S. J., – 1967 – Relationship among functional properties of Californian grassland. *Nature* (*Lond.*) 216: 168–169.

Marlange, M., – 1971 – Caractères écologiques généraux du Chaco argentin. C.N.R.S. doc. no. 58. Centre d'Etudes phytosociologiques et écologiques, Montpellier. 172 pp.

Marques Vierra, L. & J. A. Gomide, – 1968 – Composição química e produção forrageira de três variedades de capim-elefante. *Revista Ceres* 15: 245–60.
Moffat, J. K., – 1966 – Pasture establishment in the beef regions of northern Santa Fé and Corrientes, Argentina. In 'Proceedings of Ninth International Grassland Congress.' pp. 287–289. Departamento da Produção Animal da Secretaria da Agricultura do Estado de São Paulo.
Molinari, O. Garcia, – 1949 – Succession of grasses in Puerto Rico. *Rev. Agric. Puerto Rico* 39: 199–217.
Molinari, O. Garcia, – 1952 – Pasture and forage grass species and their ecology in the Caribbean area. Paper presented at Sixth International Grassland Congress. 18 pp.
Morello, J. H., (with J. Adamoli), – 1968 – La vegetación de la Republica Argentina: Las grandes unidades de vegetación y ambiente del Chaco Argentino. 1. Objectivos y Metodología. Instituto de Botanica Agricola, Ser. Fitogeográfica no. 10. Buenos Aires. 125 pp.
Morello, J. H., – 1970 – Modelo de relaciones entre pastizales y leñosas colonizadoras en el Chaco Argentino: (Plan Ecología, Difusión y Control del Vinal). *Idia* no. 276: 31–52.
Morello, J. H., N. E. Crudeli & M. Saraceno – 1971 – Los Vinalares de Formosa (Republica Argentina). (La Colonizadora Leñosa *Prosopis ruscifolia* Gris). Instituto Nacional de Tecnología Agropecuaria, Serie Fitogeográfica no. 11. Buenos Aires. 111 pp.
Morello, J. H. & Saravia Toledo, C., – 1959 – El bosque chaqueño. 1. Paisaje primitivo, paisaje natural y paisaje cultural en el oriente de Salta. 2. La ganaderia y el bosque en el oriente de Salta. *Revista Agronomica del Noroeste Argentino* 3: 5–81 and 209–258.
Myers, J. G., – 1936 – Savanna and forest vegetation of the interior Guiana plateau. *J. Ecol.* 24: 162–183.
Norris, D. O., – 1969 – Observations on the nodulation status of rainforest leguminous species in Amazonia and Guyana. *Trop. Agric. Trin.* 46: 145–151.
Oakes, A. J., – 1965 – *Digitaria* collections from South Africa. *Trop. Agric. Trin.* 42: 323–331.
Oakes, A. J., – 1968 – Replacing of hurricane grass in pastures of the dry tropics. *Trop. Agric. Trin.* 45: 235–241.
Oakes, A. J., – 1970 – Legumes in the U.S. Virgin Islands. *Turrialba* 20: 153–165.
Osbourn, D. F., – 1969 – The introduction of Pangola grass into the Caribbean Islands. *J. Brit. Grassland Soc.* 24: 76–80.
Parsons, J. J., – 1970 – The 'Africanization' of the New World tropical grasslands. *Tübinger Geographische Studien* 34: 141–153.
Parsons, J. J,. – 1972 – Spread of African pasture grasses to the American tropics. *J. Range Management* 25: 12–17.
Paula, R. R., J. A. Gomide, D. Sykes & J. R. Chaves, – 1969 – Influência de diferentes sistemas de corte sôbre o desenvolvimento radicular do capim-Gordura (*Melinis minutiflora* Beauv.). *Revta. Ceres* 16: 10–18.
Pennington, T. D. & J. Sarukhan, – 1968 – Arboles tropicales de México. Instituto Nacional de Investigaciones Forestales, Mexico/FAO, Rome. 413 pp.
Ramia, M., – 1957 – Pastos de los Llanos de Barinas. *Bol. Soc. Venez. Cienc. Natur.* 87: 271–311.
Ramia, M., – 1958 – Los médanos de Guárico occidental. *Bol. Soc. Venez. Clenc. Natur.* 91: 41–53.
Ramia, M., – 1961 – Sabanas llaneras. *El Farol* no. 197: (unpaginated).
Ramia, M., – 1964 – Distribución de Sabanas en Venezuela. *Rev. Venez. Geogr.* 4: 25–34.

Ramia, M., – 1968 – Nuestras sabanas. *El Farol* 24: 8–12.

Richards, J. A., – 1970 – Productivity of tropical pastures in the Caribbean. In 'Proceedings Eleventh International Grassland Congress,' pp. A49–56. University of Queensland, St. Lucia.

Rocha, G. Leme da, – 1972 – Trabalho apresentado à reunião especial de FAO sôbre pastos e forragens na América Tropical. 48 pp. (67 references).

Rodriguez, C., – 1970 – Efecto de dos fuentes de nitrogeno sobre el rendimiento y composición quimica de los pastos Pangola, A-24 y SR-954. *Agronomía Tropical* 20: 119–124.

Rodriguez, C. & E. de Blanco, – 1970 – Composición quimica de hojas y tallos de 21 cultivares de Elefante (*Pennisetum purpureum*, Schumacher). *Agronomía Tropical* 20: 383–396.

Rosengurtt, B., – 1944 – Estudios sobre praderas naturales del Uruguay. 4a. Las formaciones campestras y herbaceas del Uruguay. *Agros* no. 134: 1–45.

Rosengurtt, B., J. P. Gallinal, L. Bergalli, L. Aragone & E. F. Campal, – 1939 – Estudios sobre praderas naturales del Uruguay. La variabilidad de la composicíon de las praderas. *Rev. Asoc. Ing. Agron.* 11(3): 28–33.

Roseveare, G. M., – 1948 – The Grasslands of Latin America. Imperial Bureau of Pastures and Field Crops. Bulletin no. 36. 291 pp. Publisher now: Commonwealth Agricultural Bureaux, Farnham Royal, Slough.

Salette, J. E., – 1972 – Informes regionales sobre problemas, actividades y programas crecientes de desarrollo en el campo de pastos y forrajes tropicales. Región del Caribe. FAO/CIAT-CPAT/11. 14 pp.

Salette, J. & Y. Dumas, – 1970 – Constantes de comportement de *Digitaria decumbens* Stent. Relation entre la teneur en azote et le rapport feuilles/tiges; influence de différentes conditions de milieu. *C.R. hebd. Séanc. Acad. Sci. Paris* (d) 271: 200–202. Relation entre les teneurs en azote et en matière sèche: influence de différentes conditions de milieu. *C.R. hebd. Séanc. Acad. Sci. Paris* (d) 271: 324–326.

Santiago, A. A., – 1970 – Pecuária de Corte no Brasil Central. Instituto de Zootecnia, Agua Branca. 635 pp.

Schwabe, G. W., – 1968 – Towards an ecological characterisation of the South American Continent. In 'Biogeography and Ecology in South America,' vol. 1, ed. E. J. Fittkau, J. Illies, H. Klinge, G. H. Schwabe & H. Sioli, pp. 113–136. W. Junk, The Hague.

Serpa, A., – 1966 – Melhoramento da *Centrosema pubescens* Benth. 1. Obtenção de variedades de sementes permeáveis. In 'Proceedings of Ninth International Grassland Congress,' pp. 171–174. Departamento da Produção animal da Secretaria da Agricultura do Estado de São Paulo.

Shambulingappa, K. G., – 1970 – Cytological studies in *Digitaria pentzii* Stent. *J. Agric. Uni. Puerto Rico* 54: 401–408.

Sluijs, D. H. van der, – 1971 – Native grasslands of the Mesopotamia region of Argentina. *Neth. J. agric. Sci.* 19: 3–22.

Sotomayor-Ríos, A., S. C. Schank & R. Woodbury, – 1970 – Cytology and taxonomic description of two Brachiarias (Congograss and Tannergrass). *J. Agric. Uni. Puerto Rico* 54: 390–400.

Souza, D. J. A. de, – 1966 – Nodulation of indigenous Trinidad legumes. *Trop. Agric. Trin.* 43: 265–267.

Stark, J., G. K. Rutherford, J. Spector & T. A. Jones – 1959 – Soil and Land-Use Surveys no. 6. British Guiana. 1. The Rupununi Savannas (contd.) 2. The Intermediate Savannas. 3. General Remarks. Regional Research Centre of the British Caribbean at the Imperial College of Tropical Agriculture, Trinidad. 24 pp.

Tardin, A. C., C. H. Calles & J. A. Gomide, – 1971 – Desenvolvimento vegetativo do capim Guatemala. *Experientiae* 12: 1–31.

Taylor, B. W., – 1962 – The status and development of the Nicaraguan pine savannas. *Caribbean Forester* 23: 21–26.

Tergas, I. E., W. G. Blue & J. E. Moore, – 1971 – Nutritive value of fertilized Jaragua grass (*Hyparrhenia rufa*) in the wet/dry Pacific region of Costa Rica. *Trop. Agric. Trin.* 48: 1–8.

Teunissen, P. A. & J. T. Wildschut – (n.d.) – Vegetation and flora of the savannas in the Brinckheuvel Nature Reserve, Northern Suriname. Verhandelingen der Koninklijke Nederlandse Akademie van Wetenschappen, afd. Natuurkunde, 59: no. 2. 60 pp.

Thomas, C. A. & L. E. McLaren, – 1971 – Some studies on digestibility of Pangola grass (*Digitaria decumbens*) in Jamaica. *Trop. Agric. Trin.* 48: 225–235.

Troll, C. (ed.) – 1968a – Geo-Ecology of the Mountainous Regions of the Tropical Americas. Proceedings of the Unesco Mexico Symposium, 1966. Colloquium Geographicum, Geographischen Institut der Universität Bonn, 222 pp.

Troll, C., – 1968b – Das Pampaproblem in landschaftsökologischer Sicht. *Erdkunde* 22: 152–155.

Voronov, A. G., – 1970 – The effect of geographical factors on the distribution of vegetation in tropical countries. *Forestry Abst.* 31: 446–452.

Walter, H., – 1967a – Das Pampaproblem in Vergleichend Okologischer Betrachtung und seine Lösung. *Erdkunde* 21: 181-203.

Walter, H., – 1967b – The Pampa problem and its solution. In "Proceedings of Second International Seminar on Integrated Surveys of Natural Grazing Lands," pp. 3-18. ITC-UNESCO Centre, Delft. (now Enschede).

West, S. H., R. H. Biggs & J. M. Baskin, – 1968 – Growth and photosynthesis of Pangola grass, *Digitaria decumbens* Stent., in a gradient of temperatures. *Proc. Soil Crop Sci. Soc. Fla* 28: 29-35.

Whyte, R. O., – 1959 – Report of Fourth FAO Inter-American Meeting on Livestock Production, 1958, Kingston, Jamaica.

Wilson, P. N., – 1959 – Animal husbandry investigations at I.C.T.A. with special reference to the management of cattle on Pangola pastures. *J. Agric. Soc. Trin. Tobago*, Society paper no. 879. 12 pp.

ANDERSON, J. A. R., – 1971 – Personal communication.

ANDREW, W. D. & A. B. P. JAYAWARDANA, – 1970 – Development of improved pasture on patana soils in the wet-montane zone of Ceylon. *Trop. Agric. Colombo*: 126: 107–120.

APPADURAI, R. R. & R. ARASARATNAM, – 1969 – The effect of large applications of urea nitrogen on the growth and yield of an established pasture of *Brachiaria brizantha* Stapf. *Trop. Agric. Trin.* 46: 153–158.

BALACHANDRAN, N., – 1969 – Fodder production in Malaysia with particular reference to Napier (*Pennisetum purpureum*). *Kajian Veterinaire* 2: 83–88.

BHATNAGAR, H. P., – 1960 – Succession in some *Shorea robusta* forests of U.P. *J. Ind. Bot. Soc.* 39: 22–29.

BHATTEE, S. S. & C. J. THAMPI, – 1963 – Some important grasses of the Andaman Islands. *Indian For.* 89: 223–230.

BLAGDEN, C. O. & W. W. SKEAT, – 1906 – Pagan Races of the Malay Peninsula. Vols. I and II: Macmillan, London. 627 & 716 pp.

BLASCO, F., – 1971 – Montagnes du Sud de l'Inde. *Inst. fr. Pondichéry, Trav. Sec. sci. tech.* 10: 1–436.

BOR, N. L., – 1960 – Grasses of Burma, Ceylon, India and Pakistan (except Bambuseae). Pergamon Press, Oxford, London, New York, Paris. 767 pp.

BOR, N. L., – 1968 – Flora of Iraq, Vol. 9. Gramineae. Ministry of Agriculture, Baghdad. 588 pp.

BROWN, W. H., – 1919 – Vegetation of Philippine Mountains: the Relation between the Environment and Physical Types at different Altitudes. Bureau of Science, Manila. 434 pp.

BURGESS, P. F., – 1972 – Personal communication.

BURKILL, I. H. & R. E. HOLTTUM, – 1923/5 – A botanical reconnaissance upon the main range of the Peninsula at Fraser Hill. *The Gardens Bulletin: Straits Settlements* 3: 19–110.

CHAKRAVARTY, A. K., – 1968 – Role of some perennial grasses of western Rajasthan in development of grasslands of the arid zone of India. In 'Proceedings of Symposium on Recent Advances in Tropical Ecology,' pp. 620–630.

CHAKRAVARTY, A. K. & G. N. BHATI, – 1968 – Selection of grasses and legumes for pastures of the arid and semi-arid zones. 1. Variations of morphological and physiological characters in the different strains of *Cenchrus setigerus* and selection of a promising strain for forage production. *Indian For.* 94: 667–674.

CHAKRAVARTY, A. K. & R. B. DAS, – 1965 – Polymorphism in *Cenchrus ciliaris* Linn. *Ann. Arid Zone* 4: 10–16.

CHAKRAVARTY, A. K. & N. L. KACKAR, – 1967 – Effect of spacing and cutting on root development of some promising desert grasses and legumes. *J. Ind. Bot. Soc.* 46: 264–269.

CHAKRAVARTY, A. K. & N. L. KACKAR, – 1970 – Selection of grasses and legumes for pasture of the arid and semi-arid zones. 2. Variation of morphological and physiological characters in *Lasiurus sindicus* Henr. and selection of promising type for forage production. *Indian For.* 96: 437–443.

CHAKRAVARTY, A. K. & N. L. KACKAR, – 1971 – Selection of grasses and legumes for pasture of the arid and semi-arid zones. 4. *Panicum antidotale* Retz. Selection of a promising type. *Indian For.* 97: 509–516.

CHAKRAVARTY, A. K. & L. KALKANI, – 1966 – Study on variation in seed yielding components of *Cenchrus ciliaris* Linn. *Ann. Arid Zone* 5: 65–71.

CHAKRAVARTY, A. K., R. RATAN & K. MURARI, – 1970 – Variation in morphological and physiological characters in bunch grass (*Cenchrus ciliaris* L.) and selection of high-yielding, nutritious types. *Indian J. agric. Sci.* 40: 912–916.

CHAMPION, H. G. & S. K. SETH, – 1968 – A Revised Survey of the Forest Types of India. Government of India, Manager of Publications, Delhi. 404 pp.

CHATTERJEE, B. N. & A. KUMAR, – 1964 – Inter-strain variations in *Pennisetum pedicellatum*. *Indian For.* 90: 477–483.

CHATTERJEE, B. N. PREMCHAND & R. D. SINGH, – 1969 – Herbage growth in spear grass (*Heteropogon contortus*) swards mixed with legumes in the Indian tropics. *Ranchi Uni. J. agric. Res.* 4: 5–7.

CHENG, S. H., – 1969 – The rice industry of Malaya – a historical survey. *J. Malayan Branch Royal Asiatic Soc.* 42 (2): 130–144.

COCKBURN, P. F., – 1969 – Personal communication.

CORNER, E. J. H., – 1949 – The durian theory or the origin of the modern tree. *Ann. Bot.* 13: 367–414.

D'CRUZ, R. & P. S. REDDY, – 1968 – Apomixis in *Pennisetum massaicum* Stapf. *Science and Culture* 34: 255–257.

D'CRUZ, R. & P. S. REDDY, – 1969 – Note on relative forage yield and chemical composition of species and hybrids of *Dichanthium* species. *Indian J. agric. Sci.* 39: 439–440.

DABADGHAO, P. M., S. P. MARWAHA, B. S. GUPTA, R. B. DAS & R. DEB ROY, – 1963 – Root ecology of some promising desert grasses of Rajasthan. *Ann. Arid Zone* 1: 163–173.

DABADGHAO, P. M. & K. A. SHANKARNARAYAN, – 1970 – Studies of *Iseilema*, *Sehima* and *Heteropogon* communities of the *Sehima-Dichanthium* zone. In 'Proceedings of Eleventh International Grassland Congress,' pp. 36–38. University of Queensland Press, St. Lucia.

DABADGHAO, P. M. & K. A. SHANKARNARAYAN, – 1972 – The Grass Cover of India. Indian Council of Agricultural Research Scientific Monograph, New Delhi.

DAFTARDAR, S. Y. & G. K. ZENDE, – 1968 – Periodical changes in the protein contents of Gajraj grass. *Poona agric. Coll. Mag.* 58: 110–116.

DAVIES, J. G., – 1970 – Pasture development in the sub-tropics, with special reference to Taiwan. *Trop. Grasslands* 4: 7–16.

GAUSSEN, H., P. LEGRIS & F. BLASCO, – 1967 – Bioclimats du Sud-Est Asiatique. *Inst. fr. Pondichéry, Trav. Sec. Sci. tech.* 3(4): 1–115.

GILLILAND, H. B., – 1971 – An Illustrated Systematic Account of the Malayan Flora, Including Commonly Cultivated Plants: Vol. 3. Grasses of Malaya. With contributions by R. E. HOLTTUM & N. L. BOR, edited by H. M. BURKILL. Government Printer, Singapore. 319 pp.

GROOME, J. ST. J., – 1972 – Personal communication.

GUPTA, R. K. & P. C. NANDA, – 1970 – Grassland types and their ecological succession in the western Himalayas. In 'Proceedings of Eleventh International Grassland Congress,' pp. 10–13. University of Queensland Press, St. Lucia.

GUPTA, S. K., – 1972 – Energy structure of standing crop in certain grasslands at Gyanpur. *Tropical Ecol.* 13: 89–95.

HAN, I. K., D. A. KIM & S. H. PARK, – 1970 – Seasonal changes of chemical composition and of dry matter digestibility of Korean native herbage plants. In 'Proceedings of Eleventh International Grassland Congress,' pp. 92–95. University of Queensland Press, St. Lucia.

HENDERSON, M. R., – 1939 – The flora of the limestone hills of the Malay Peninsula. *J. Malayan Branch Roy. Asiatic Soc.* 17: 13–87.

HOFSTRA, J. J., S. AKSORNKOAE, S. ATMOWIDJOJO, J. F. BANAAG, SANTOSA, R. A. SASTROHOETOMO & L. T. N. THU, – 1972 – A study on the occurrence of plants with a low CO_2 compensation point in different habitats in the tropics. *Annales Bogorienses* 5: 143–157.

HSU, C. C., – 1971 – A guide to the Taiwan grasses, with keys to subfamilies. *Taiwania* 16: 199–341.

JAIN, S. K., – 1967 – The genus *Cynodon* Rich. ex Pers. in India. *Bull. bot. Surv. India* 9: 134–151.

JAUHAR, P. P., – 1968 – Genome relationships in two species of *Pennisetum*. In 'Proceedings Twelfth International Congress of Genetics,' p. 183.

JAUHAR, P. P. & A. B. JOSHI, – 1966 – Cytotaxonomic investigations in the *Panicum maximum* complex. 1. Morphological studies. *Bull. bot. Surv. India* 8: 287–295.

JAUHAR, P. P. & A. B. JOSHI, – 1969 – Cytotaxonomic investigations in the *Panicum maximum* complex. 2. Cytological features, chromosome diminution and evolution of the karyotype. *Cytologia* 34: 222–233.

JAVIER, E. Q., – 1970a – The flowering habits and mode of reproduction of Guinea grass (*Panicum maximum* Jacq.). In 'Proceedings Eleventh International Grassland Congress,' pp. 284–289. University of Queensland Press, St. Lucia.

JAVIER, E. Q., – 1970b – The phenology, flowering habits and mode of reproduction of *Pennisetum purpureum* Schumach., *Brachiaria mutica* (Forsk.) Stapf, and *Panicum maximum* Jacq. *Diss. Abstr. Int.* 30(9) 3949B–50B.

JOHNSTON, A. & E. HUSSAIN, – 1963 – Grass cover types of West Pakistan. *Pak. J. For.* 13: 239–247.

KAUL, R. N. & B. N. GANGULI, – 1963 – Fodder potential of *Zizyphus* in the scrub grazing lands of arid zone. *Indian For.* 89: 623–630.

KELLMAN, M. C., – 1969 – Shifting cultivation in upland Mindanao. *J. Trop. Geogr.* 28: 40–56.

KENG, Yi-Li, – 1965 – Flora Illustrata Plantarum Primarum Sinicarum: Gramineae. Nanking University/Academia Sinica, Peking. 1178 pp. (in Chinese).

KHAN, C. M. ANWAR, – 1970 – Range management – a challenge in West Pakistan. *Pak. J. For.* 20: 329–350.

KHAN, C. M. ANWAR, – 1971a – Rainfall pattern and monthly forage yields in Thal ranges of Pakistan. *Pak. J. For.* 21: 93–102.

KHAN, C. M. ANWAR, – 1971b – Range management strategy in Pakistan. 1. *Pak. J. For.* 21: 313–324.

KHYBRI, M. L. & D. D. MISHRA, – 1967 – Root studies on some selected grasses in eastern Nepal. *Indian For.* 93: 400–406.

LABROUE, L., P. LEGRIS & M. VIART, – 1965 – Bioclimats du sous-continent indien. *Inst. fr. Pondichéry, Trav. Sect. scien. techn.* 3(3): 1–32.

LAZARIDES, M., – 1971 – Personal communication.

LEE, YONG NO, – 1966 – Manual of the Korean Grasses. Ewha Womans University Press, Seoul. 300 pp.

LEGRIS, P. & M. VIART, – 1961 – Bioclimates of South-India and Ceylon. *Inst. fr. Pondichéry, Trav. Sect. scient. tech.* 3(2): 165–179.

LIU, T., – 1968, 1970, 1971a, 1971b – Studies on the classification of the climax vegetation communities of Taiwan. 1. *Taiwan Forestry Res. Inst. Bull.* no. 166. 25 pp. 2. *Taiwan Forestry Res. Inst. Bull.* no. 203. 24 pp. 3. *Proc. Nat. Sci. Council Taiwan* no. 4: 1–36. 4. *J. Agric. Assoc. China n.s.* no. 76: 39–62. (In Chinese, with English summaries).

MAHARASHTRA STATE, INDIA, – 1969 – Review of Research on Grasslands, Forages and Fodder Crops of Maharashtra State. Department of Agriculture, Research Bulletin no. 15. M. S. Poona. 185 pp.

MEHER-HOMJI, V. M., – 1963 – Les bioclimats du sub-continent indien et leurs types analogues dans le monde. *Inst. fr. Pondichéry, Trav. Sec. sci. techn.* 7(1–2): 1–386 pp.

MEHER-HOMJI, V. M., – 1964 – Life-forms and biological spectra as epharmonic criteria of aridity and humidity in the tropics. *J. Indian Bot. Soc.* 43: 424–430.

MEHRA, K. L., – 1961 – Chromosome numbers, geographical distribution and taxonomy of the *Dichanthium annulatum* complex. *Phyton* 17: 157–166.

MEHRA, K. L., – 1962 – The *Dichanthium annulatum* complex. 1. Morphology. *Phyton* 18: 87–94.

MEHRA, K. L. & A. P. SINGH – 1968 – Chromosome manipulation and genotypic control of chromosome behaviour in *Dichanthium annulatum* complex and hybrids. *Nucleus* Suppl. 372–382.

MEIJER, W., – 1965 – A botanical guide to the flora of Mount Kinabalu. In 'Symposium on Ecological Research in Humid Tropics Vegetation,' pp. 325–364. Government of Sarawak/UNESCO Science Cooperation Office for Southeast Asia.

MERRILL, E. D., – 1906 – An enumeration of Philippine Gramineae, with keys to genera and species. *Philippine Agricultural J.* 1:

MISHRA, M. L. & B. N. CHATTERJEE, – 1968 – Seed production in the forage grasses *Pennisetum polystachyon* and *Andropogon gayanus* in the Indian tropics. *Trop. Grasslands* 2: 51–56.

MISRA, R., – 1969 – Studies on the primary productivity of terrestrial communities at Varanasi. *Trop. Ecol.* 10: 1–15.

MONOD DE FROIDEVILLE, C., – 1968 – Poaceae (Gramineae). In 'Flora of Java', vol. III, eds. C. A. BACKER & R. C. BAKHUIZEN VAN DEN BRINK, JR., pp. 495–641. N.V. P. Noordhoff, Groningen.

MUELLER-DOMBOIS, D., – 1969 – Ecogeographic analysis of a climate map of Ceylon with particular reference to vegetation. *Ceylon For.* 8: (3 & 4): 1–2.

NODA, M., – 1972 – Flora of the North-Eastern Province (Manchuria) of China (in Japanese). Kazama Bookshop Co., Tokyo. 1613 pp.

OKE, J. G., – 1968 – Studies in Forage Grasses of major Importance in the Maharashtra State. Doctoral Thesis, Poona. 501 pp.

OKE, J. G., – 1971 – Some new improved strains of pasture grasses for Maharashtra State. *Indian For.* 97: 654–657.

OKE, J. G., – 1972 – Studies in grassland ecology of the Maharashtra State. 1. Ecological classification of grassland patterns found in different ecological habitats and their botanical characterization. *Indian For.* 98: 86–106.

PANDEYA, S. C., – 1953 – Studies on the morphology and ecology of three species of *Dichanthium* Willemet. *J. Indian bot. Soc.* 32: 86–100.

PANDEYA, S. C., – 1961 – On some new concepts in phytosociological studies of grasslands. 1. Dominance diagrams. 2. Community coefficient (F×C)ICC. *J. Indian bot. Soc.* 40: 263–266; 267–270.

PANDEYA, S. C., – 1969 – Ecology of grasslands of Sagar, Madhya Pradesh. 3. Edaphic factors in the distribution of grassland associations. *Trop. Ecol.* 10: 163–182.

PANDEYA, S. C. & P. K. JAYAN, – 1970 – Population differences in buffel grass (*Cenchrus ciliaris*) of Ahmedabad, India: productivity under various agronomic conditions. In 'Proceedings of Eleventh International Grassland Congress,' pp. 239–244. University of Queensland Press, St. Lucia.

PATEL, R. I., – 1969 – Personal communication.

PATIL, B. D. & A. SINGH, – 1963 – Studies on anjan grass. 1. Intervarietal variation in forage and seed production. *Indian J. agric. Sci.* 33: 44–51.

PATIL, B. D. , A. SINGH & R. GHOSH, – 1964 – Intervarietal variation in persistency and aggressiveness in anjan grass (*Cenchrus setigerus* Vahl.) and the technique for its estimation. *Indian For.* 90: 40–49.

PELZER, K. J., – 1968 – Man's role in changing the landscape of southeast Asia. *J. Asian Studies* 27: 269–279.

RAI, R. R., G. S. SINGH & S. N. SINGH, – 1968 – Studies on Para grass (*Panicum barbinode* or *Brachiaria mutica* Stapf). 2. Effect of hoeing, organic and inorganic nitrogen on the chemical composition and yield of para grass. *Balwant Vidyapeeth J. Agric. Sci. Res.* 8: 26–35.

RAO, C. K., – 1972 – Angiosperm genera endemic to the Indian floristic region and its neighbouring areas. *Ind. For.* 98: 560–566.

RAO, S. M. & C. RAMALINGAM, – 1964 – Studies on the flowering, seed formation and seed viability of Kikuyu grass (*Pennisetum clandestinum* Hochst.). *Madras agric. J.* 51: 431–434.

RATTRAY, J. M., – 1969 – Personal communication

REDDY, P. S. & R. D'CRUZ, – 1968 – Cytomorphological studies in the interspecific hybrids of *Dichanthium* (Willemet). 2. *D. annulatum* (Forssk) Stapf × D. *caricosum* (Linn.) A. Camus. *J. Ind. Bot. Soc.* 47: 271–277.

SANTIAGO, A. A., – 1966 – Studies in autecology of *Imperata cylindrica* (L.) Beauv. In 'Proceedings of Ninth International Grassland Congress,' pp. 499–502. Departamento da Produção Animal da Secretaria da Agricultura do Estado de São Paulo.

SHANKARNARAYAN, K. A., – 1963 – Parallelism between grassland and forest types of Bellary District. *Ann. Arid Zone* 1: 132–141.

SHARMA, B. M., – 1968 – Root systems of some desert plants at Churu, Rajasthan. *Indian For.* 94: 240–246.

SINGH, A. & B. D. PATIL, – 1970 – Studies on induced polyploids of forages. 1. Some aspects of growth, morphology and seeding habit of tetraploid Siratro. *Indian J. Sci. and Ind.* 4: 43–48.

SINGH, J. S. & R. MISRA, – 1969 – Diversity, dominance, stability and net production in the grasslands at Varanasi, India. *Canad. J. Bot.* 47: 425–427.

SINGH, M. & N. D. ARORA, – 1970 – Introduction and evaluation of *Pennisetum pedicellatum* as summer fodder. In 'Proceedings Eleventh International Grassland Congress,' pp. 621–624. University of Queensland Press, St. Lucia.

STAPF, O., – 1894 – On the flora of Mount Kinabalu in North Borneo. *Trans. Linn. Soc., London, 2nd Ser. Botany* 4: 69–263.

STEENIS, C. G. G. J. VAN & C. MONOD DE FROIDEVILLE, – 1968 – Malesia. In 'Grasslands of the Monsoon,' by R. O. WHYTE, pp. 224–243. Faber and Faber, London/ Praeger, New York.

STODDART, D. R. & M. YONGE, (eds.), – 1971 – Regional Variation in Indian Ocean Coral Reefs. Academic Press, New York. 584 pp.

SUBRAHMANYAM, V. P. & C. V. S. SASTRY, – 1969 – Climatic analogues for the dry zone of India. *Trop. Ecol.* 10: 34–44.

SUNDARARAJ, D. D. & D. MEENAKSHISUNDARAM, – 1955 – Recent advances in agriculture – plant introduction and its further possibilities. *Madras Agric. J.* 42: 133–135.

THOMAS, R. & L. R. HUMPHREYS, – 1970 – Pasture improvement at Na Pheng, central Laos. *Trop. Grasslands* 4: 229–236.

TROLL, C. (ed.), – 1972 – Geoecology of the high-mountain regions of Eurasia. *Erdwissenschaftliche Forschung* 4: pp. 1–16; 237–243; 262–263; 264–275.

VERBOOM, W. C., – 1968 – Grassland successions and associations in Pahang, central Malaya. *Trop. Agric. Trin.* 45: 47–59.

VERSTAPPEN, H. T., – 1969 – Aerial survey and rural development in south Asia. In 'Jahrbuch des Südasiens-Institut der Universität Heidelberg 1968/69,' pp. 1–6. Otto Harrossowitz, Wiesbaden.

WALEY, A., – 1937 – Three Ways of Thought in Ancient China. Allen and Unwin, London. 245 pp.

WANG, CHI-CHU, – 1959 – A comparison of the distribution of the natural grasses and legumes in climatic regions of Florida, USA, and northern Taiwan. *Natural Sciences and Agriculture Society* 7: 67–76.

WANG, CHI-WU, – 1961 – The Forests of China. Maria Moors Cabot Foundation Publication Series no. 5. 313 pp.

WHITFORD, H. N., – 1906 – The vegetation of the Lamao Forest Reserve. *Philippines J. Science* 1: 373

WHYTE, R. O., – 1964 – The Grassland and Fodder Resources of India (2nd ed.). Indian Council of Agricultural Research, Scientific Monograph no. 22, New Delhi. 553 pp.

WHYTE, R. O., – 1968b – Land, Livestock and Human Nutrition in India. F. T. Praeger, New York. 309 pp.

WHYTE, R. O., – 1971 – Grazing in the land ecosystems of India. *Ann. Arid Zone* 10: 111–119.

WHYTE, R. O., – 1972a – The nature and utilization of the grazing resources. In 'Nomads and Pastoralists in South Asia,' Südasien-Institut, University of Heidelberg.

WHYTE, R. O., – 1972b – The Gramineae, wild and cultivated, of monsoonal and equatorial Asia. 1. Southeast Asia. *Asian Perspectives*, 15:

WHYTE, R. O., – 1973 – Grasses and grasslands. In 'UNESCO Review of Research: The Natural Resources of Humid Tropical Asia,' UNESCO, Paris.

WRIGHT, R. L., – 1964 – Land system survey of the Nagarparkar Peninsula, Pakistan. *Arid Zone* (*UNESCO*), no. 25: 5–13.

Australia and Pacific

ANDREW, C. S., – 1963 – Copper deficiency symptoms of some tropical and temperate pasture legumes. *Austral. J. Agric. Res.* 14: 654–659.

ANDREW, C. S., – 1965 – Mineral nutrition of sub-tropical pasture species. *J. Austral. Inst. Agric. Sci.* 31: 3–10.

ANDREW, C. S., – 1966 – A kinetic study of phosphate absorption by excised roots of *Stylosanthes humilis. Phaseolus lathyroides, Desmodium uncinatum, Medicago sativa* and *Hordeum vulgare. Austral. J. Agric. Res.* 17: 611–624.

ANDREW, C. S. & W. H. J. PIETERS, – 1970a – Manganese toxicity symptoms of one temperate and seven tropical pasture legumes. CSIRO Division of Tropical Pastures Technical Paper no. 4. 8 pp.

ANDREW, C. S. & W. H. J. PIETERS, – 1970b – Effect of potassium on the growth and chemical composition of some pasture legumes. 3. Deficiency symptoms of ten tropical pasture legumes. Division of Tropical Pastures Technical Paper no. 5. 11 pp.

ANDREW, C. S. & M. F. ROBINS, – 1969a – The effect of phosphorus on the growth and chemical composition of some tropical pasture legumes. 1. Growth and critical percentages of phosphorus. 2. Nitrogen, calcium, magnesium, potassium and sodium contents. *Austral. J. Agric. Res.* 20: 665–672, 675–685.

ANDREW, C. S. & M. F. ROBINS, – 1969b – The effect of potassium on the growth and chemical composition of some tropical and temperate pasture legumes. 1. Growth and critical percentages of potassium. 2. Potassium, calcium, magnesium, sodium, nitrogen, phosphorus and chloride. *Austral. J. Agric. Res.* 20: 999–1007, 1009–1021.

BLAKE, J. D. & G. N. RICHARDS, – 1970 – Polysaccharides of tropical pasture herbage. 1. Studies on the distribution of the major polysaccharide components of spear grass (*Heteropogon contortus*) during growth. 2. A xylan from the leaf of spear grass. *Austral. J. Chem.* 23: 2353–2360, 2361–2368.

BOWERS, N., – 1969 – The ascending grasslands: an anthropological study of ecological succession in a high mountain valley of New Guinea. *Diss. Abstr.* 29(8) 2718B.

BRAY, R. A., – 1970 – Development of grazing lucerne for subtropical conditions. In 'Proceedings of Eleventh International Grassland Congress,' pp. 261–263. University of Queensland Press, St. Lucia.

BRAY, R. A., J. B. HACKER & D. E. BYTH, – 1969 – Root mapping of three tropical pasture species using 32P. *Austral. J. Exp. Agric. Anim. Husb.* 9: 445–448.

BRYAN, W. W., – 1961 – *Lotononis bainesii* Baker: a legume for subtropical pastures. *Austral. J. Exp. Agric. Anim. Husb.* 1: 4–10.

BRYAN, W. W., – 1963 – The search for tropical pasture legumes – a progress report. *J. Agric. Sci.* 29: 149–153.

BRYAN, W. W., – 1966 – The pasture value of species of *Desmodium.* In 'Proceedings of Tenth International Grassland Congress,' pp. 311–315.

BRYAN, W. W., – 1969 – *Desmodium intortum* and *Desmodium uncinatum. Herbage Abstracts* 39: 183–91. (95 references).

BRYAN, W. W., J. P. SHARPE & K. P. HAYDOCK, – 1971 – Some factors affecting the growth of lotononis (*Lotononis bainesii*). *Austral. J. Exp. Agric. Anim. Husb.* 11: 29–34.

BRYAN, W. W. & N. H. SHAW, – 1964 – *Paspalum plicatulum* Michx. *Austral. J. Exp. Agric. Anim. Husb.* 4: 17–21.

BURT, R. L., – 1968 – Growth and development of buffel grass (*Cenchrus ciliaris*). *Austral. J. Exp. Agric. Anim. Husb.* 8: 712–719.

BURT, R. L. & J. P. EBERSOHN, – 1969 – Some comments on pasture plant introduction in Queensland. *J. Austral. Inst. Agric. Sci.* 35: 124–127.

BURT, R. L., L. A. EDYE, B. GROF & R. J. WILLIAMS, – 1970 – Assessing the agronomic potential of the genus *Stylosanthes* in Australia. In 'Proceedings Eleventh International Grassland Congress,' pp. 219–223. University of Queensland, St. Lucia.

BURT, R. L. & K. P. HAYDOCK, – 1968 – Yield stability as an aid to the preliminary assessment of introduced forage plants. *J. Austral. Inst. Agric. Sci.* 34: 228–231.

BYRNE, G. F. & K. TOGNETTI, – 1969 – Simulation of a pasture-environment interaction. *Agric. Met.* 6: 151–163.

BYTH, D. E., – 1964 – Breeding system and chromosome number in *Lotononis bainesii* Baker. *Nature (Lond.)* 202: 830.

CAMERON, D. F., – 1967a – Flowering in Townsville lucerne (*Stylosanthes humilis*). 1. Studies in controlled environments. *Austral. J. Exper. Agric. Anim. Husb.* 7: 489–494.

CAMERON, D. F., – 1967b – 2. The effect of latitude and time of sowing on the flowering time of single plants. *Austral. J. Exper. Agric. Anim. Husb.* 7: 495–500.

CAMERON, D. F., – 1967c – Flowering time and the natural distribution and dry matter production of Townsville lucerne (*Stylosanthes humilis*) population. *Austral. J. Exper. Agric. Anim. Husb.* 7: 501–508.

CAMERON, D. F., – 1967d – Chromosome number and morphology of some introduced *Stylosanthes* species. *Austral. J. Agric. Res.* 18: 375–379.

CAMERON, D. F., – 1967e – Hardseededness and seed dormancy of Townsville lucerne (*Stylosanthes humilis*) selections. *Austral. J. Exper. Agric. Anim. Husb.* 7: 237–240.

CAMERON, D. F., – 1970 – Townsville lucerne (*Stylosanthes humilis*): a comparison of introductions from Brazil and Mexico with naturalized Australian ecotypes. In 'Proceedings of Eleventh International Grassland Congress,' pp. 184–187. University of Queensland Press, St. Lucia.

CAMPBELL, C. M., L. B. SHERROD & S. M. ISHIZAKI, – 1971 – Effects of supplemental phosphorus on utilization of late regrowth Kikuyu grass and Pangola grass nutrients. Hawaii Agricultural Experiment Station Research Report no. 192. 11 pp.

CATCHPOOLE, V. R., – 1970 – The silage fermentation of some tropical pasture plants. In 'Proceedings of the Eleventh International Grassland Congress,' pp. 891–894. University of Queensland Press, St. Lucia.

CATCHPOOLE, V. R. & E. F. HENZELL, – 1971 – Silage and silage-making from tropical herbage species. *Herbage Abstracts* 41: 213–221. (86 references).

CHRISTIE, G. S., F. G. DE MUNK, N. P. MADSEN & M. P. HEGARTY, – 1971 – The effects of an arginine antagonist on stimulated human lymphocytes in culture. *Pathology* 3: 139–144.

COALDRAKE, J. E., – 1970 – The Brigalow. In 'Australian Grasslands,' ed. R. M. MOORE, pp. 123–140. Australian National University Press, Canberra.

COLMAN, R. L. & A. LAZENBY, – 1970 – Factors affecting the response of some tropical and temperate grasses to fertilizer nitrogen. In 'Proceedings of Eleventh International Grassland Congress,' pp. 392–397. University of Queensland Press, St. Lucia.

CONNOR, D. J. & O. CARTLEDGE, – 1970 – Observed and calculated photosynthetic rates of *Chloris gayana* communities. *J. appl. Ecol.* 7: 353–362.

CULL, J. K. & J. P. EBERSOHN, – 1969 – Dynamics of semi-arid plant communities in Western Queensland. 1. Population shifts of two invaders: *Cenchrus ciliaris* cv Gayndah and *Heteropogon contortus*. *Queensland J. Agric. Anim. Sci.* 26: 193–198.

CULVENOR, C. C. J., M. C. FOSTER & M. P. HEGARTY, – 1971 – A total synthesis of indospicine, 6-Amidino-2-Aminohexanoic acid. *Austral. J. Chem.* 24: 371–375.

DONALDSON, L. E., R. I. HAMILTON, L. J. LAMBOURNE & D. A. LITTLE, – 1970 – Assessing *Leucaena leucocephala* for deleterious effects in cattle and sheep. In 'Proceedings of Eleventh International Grassland Congress,' pp. 780–821. University of Queensland Press, St. Lucia.

DOWNES, R. W., – 1969 – Differences in transpiration rates between tropical and temperate grasses under controlled conditions. *Planta* 88: 261–273.

DOWNES, R. W., – 1970 – Differences between tropical and temperate grasses in rates of photosynthesis and transpiration. In 'Proceedings of Eleventh International Grassland Congress,' pp. 527–530. University of Queensland Press, St. Lucia.

DOWNES, R. W., – 1971 – Relationship between evolutionary adaptation and gas exchange characteristics of diverse *Sorghum* taxa. *Austral. J. Biol. Sci.* 24: 843–852.

EVANS, L. T., & R. B. KNOX, – 1969 – Environmental control of reproduction in *Themeda australis*. *Austral. J. Bot.* 17: 375–389.

EVANS, T. R., – 1967 – Primary evaluation of grasses and legumes for the Northern Wallum of south-east Queensland. *Tropical Grasslands* 1: 143–152.

FITZPATRICK, E. A. & H. A. NIX, – 1970 – The climatic factor in Australian grassland ecology. In 'Australian Grasslands,' ed. R. M. MOORE, pp. 3–26. Australian National University Press, Canberra.

GARTNER, J. A., – 1969a – Effect of fertilizer nitrogen on a dense sward of kikuyu, paspalum and carpet grass. 1. Botanical composition, growth and nitrogen uptake. *Queensland J. Agric. Anim. Sci.* 26: 21–33.

GARTNER, J. A., – 1969b – 2. Interactions with phosphorus and potassium. *Queensland J. Agric. Anim. Sci.* 26: 365–372.

GARTNER, J. A. & M. L. EVERETT, – 1970 – 3. Nitrogen source. *Queensland J. Agric. Anim. Sci.* 27: 73–87.

GATES, C. T., – 1970 – Physiological aspects of the rhizobial symbiosis in *Stylosanthes humilis*, *Leucaena leucocephala* and *Phaseolus atropurpureus*. In 'Proceedings of Eleventh International Grassland Congress,' pp. 442–446. University of Queensland Press, St. Lucia.

GATES, C. T., J. R. WILSON & N. H. SHAW, – 1966 – Growth and chemical composition of Townsville lucerne (*Stylosanthes humilis*). 2. Chemical composition with special reference to cations, as affected by the principal constituent elements of molybdenized superphosphate. *Austral. J. Exper. Agric. Anim. Husb.* 6: 266–276.

GILLARD, P., – 1970 – Pasture development in the dry tropics of north Queensland. In 'Proceedings of Eleventh International Grassland Congress,' pp. 807–810. University of Queensland Press, St. Lucia.

GILLARD, P., – 1971 – *Urochloa mosambicensis* – an easily established perennial grass companion for Townsville stylo. *Tropical Grasslands* 5: 131–135.

GOULD, K. F., – 1971 – Somatic chromosome numbers in some *Digitaria* spp. Division of Tropical Pastures Technical Paper no. 7. 7 pp.

GRAHAM, T. W. G. & L. R. HUMPHREYS, – 1970 – Salinity response of cultivars of buffel grass (*Cenchrus ciliaris*). *Austral. J. Exp. Agric. Anim. Husb.* 10: 725–728.

GRAY, S. G., – 1968 – A review of research on *Leucaena leucocephala*. *Tropical Grasslands* 2: 19–30.

GRAY, S. G., – 1970 – The place of trees and shrubs as sources of forage in tropical and subtropical pastures. *Tropical Grasslands* 4: 57–62.

GROF, B., – 1968 – Viability of seed of *Brachiaria decumbens*. *Queensland J. Agric. Anim. Sci.* 25: 149–152.

GROF, B., – 1969a – Notes on selections from hybrid derivatives of elephant grass (*Pennisetum purpureum* Schum.). *Queensland J. Agric. Anim. Sci.* 26: 49–53. Elephant grass for warmer and wetter lands. *Queensland Agric. J.* 95: 227–234.

GROF, B., – 1969b – Viability of Para grass (*Brachiaria mutica*) seed and the effect of fertilizer nitrogen on seed yield. *Queensland J. Agric. Anim. Sci.* 26: 271–276.

GROF, B., – 1970 – Interspecific hybridization in *Centrosema*: hybrids between *C. brasilianum*, *C. virginianum* and *C. pubescens*. *Queensland J. Agric. Anim. Sci.* 27: 385–390.

GROF, B. & W. A. T. HARDING, – 1970 – Yield attributes of some species and ecotypes of *Centrosema* in north Queensland. *Queensland J. Agric. Anim. Sci.* 27: 237–43.

HAANTJES, H. A., J. A. MABBUTT & R. PULLEN, – 1965 – Anthropogenic grasslands in the Sepik Plains, New Guinea. *Pacific Viewpoint* 6: 215–219.

HACKER, J. B., – 1968 – Failure of chromosome pairing in a probable *Desmodium intortum* × *D. sandwicense* hybrid. *Austral. J. Bot.* 16: 545–550.

HACKER, J. B., – 1971 – Digestibility of leaves, leaf sheaths and stems in *Setaria. J. Austral. Inst. Agric. Sci.* 37: 154–155.

HACKER, J. B., – 1972 – Seasonal yield distribution in *Setaria. Austral. J. Exp. Agric. Anim. Husb.* 12: 36–42.

HACKER, J. B. & R. J. JONES, – 1969 – The *Setaria sphacelata* complex – a review. *Tropical Grasslands* 3: 13–34.

HACKER, J. B. & R. J. JONES, – 1971 – The effect of nitrogen fertilizer and row spacing on seed production in *Setaria sphacelata. Tropical Grasslands* 5: 61–73.

HEGARTY, M. P. & A. W. POUND, – 1970 – Indospicine, a hepatotoxic amino acid from *Indigofera spicata*: isolation, structure and biological studies. *Austral. J. Biol. Sci.* 23: 831–842.

HENZELL, E. F., – 1962 – Nitrogen fixation and transfer by some tropical and temperate pasture legumes in sand culture. *Austral. J. Exper. Agric. Anim. Husb.* 2: 132–140.

HENZELL, E. F., – 1967 – Tropical pasture legumes in Northern Australia. Paper to symposium on legumes in tropical agriculture. *Proc. Soil and Crop Sci. Soc. Florida* 27: 322–338. (80 references).

HENZELL, E. F., – 1968 – Sources of nitrogen for Queensland pastures. *Tropical Grasslands* 2: 1–17.

HENZELL, E. F., – 1970a – Problems in comparing the nitrogen economies of legume-based and nitrogen-fertilized pasture systems. In 'Proceedings Eleventh International Grassland Congress,' pp. A112–120. University of Queensland Press, St. Lucia.

HENZELL, E. F., – 1970b – Use of nitrogenous fertilizers on sub-tropical pastures in Queensland. *J. Austral. Institute of Agric. Sci.* 36: 206–213.

HENZELL, E. F., I. F. FERGUS & A. E. MARTIN, – 1966 – Accumulation of soil nitrogen and carbon under a *Desmodium uncinatum* pasture. *Austral. J. Exper. Agric. Anim. Husb.* 6: 157–160.

HENZELL, E. F., A. E. MARTIN & P. J. ROSS, – 1970 – Recovery of fertilizer nitrogen by Rhodes grass. In 'Proceedings of Eleventh International Grassland Congress,' pp. 411–414. University of Queensland Press, St. Lucia.

HENZELL, E. F. & D. O. NORRIS, – 1962 – Processes by which nitrogen is added to the soil/plant system. In 'A review of nitrogen in the tropics with particular reference to pastures,' pp. 1–18. Bulletin no. 46, Commonwealth Bureau of Pastures and Field Crops, Hurley.

HENZELL, E. F. & D. J. OXENHAM, – 1964 – Seasonal changes in the nitrogen content of three warm-climate pasture grasses. *Austral. J. Exper. Agric. Anim. Husb.* 4: 336–344.

HEYLIGERS, P. C., – 1966 – Observations on *Themeda australis* – *Eucalyptus* savannah in Papua. *Pacific Science* 20: 477–489.

HEYLIGERS, P. C., – 1972 – Analysis of the plant geography of the semi-deciduous scrub and forest and the Eucalypt savannah near Port Moresby. *Pacific Science* 26: 229–241.

HUGH, E., – 1968 – Grasses and legumes in coconut plantations of some South Pacific Commission Territories. Paper to Third Session of FAO Technical Working Party on Coconut Production, Protection and Processing, PL CNP/85/5. 8 pp.

HUNTER, R. A., B. L. MCINTYRE, & R. J. MCILROY, – 1970 – Water-soluble carbohydrates of tropical pasture grasses and legumes. *J. Sci. Food Agric.* 21: 400–405.

HUTTON, E. M., – 1962 – Siratro – a tropical pasture legume bred from *Phaseolus atropurpureus*. *Austral. J. Exp. Agric. Anim. Husb.* 2: 117–125.

HUTTON, E. M., – 1965 – A review of the breeding of legumes for tropical pastures. *J. Austral. Inst. Agric. Sci.* 31: 102–109.

HUTTON, E. M., – 1968 – Australia's pasture legumes. *J. Austral. Inst. Agric. Sci.* 34: 203–218 (84 references).

HUTTON, E. M., – 1971a – Plant improvement for increased animal production. *J. Austral. Inst. Agric. Sci.* 37: 212–225.

HUTTON, E. M., – 1971b – Variation in salt response between tropical pasture legumes. *SABRAO Newsletter* 3: 75–81.

HUTTON, E. M. & L. B. BEALL, – 1971 – Flower and seed colours as markers for new derivatives of Siratro (*Phaseolus atropurpureus*). *Tropical Grasslands* 5: 99–103.

ISBELL, R. F., – 1969 – Distribution of black spear grass (*Heteropogon contortus*) in tropical Queensland. *Tropical Grasslands* 3: 35–41.

JOHNS, R. J. & P. F. STEVENS, – 1971 – Mount Wilhelm Flora: A Check List of the Species. Botany Bulletin no. 6, Department of Forests, New Guinea. 61 pp.

JONES, R. J., – 1967 – Effects of close cutting and nitrogen fertilizer on growth of a Siratro (*Phaseolus atropurpureus*) pasture at Samford, south-eastern Queensland. *Austral. J. Exper. Agric. Anim. Husb.* 7: 157–161.

JONES, R. J., J. G. DAVIES & R. B. WAITE, – 1967 – The contribution of some tropical legumes to pasture yields of dry matter and nitrogen at Samford, south-eastern Queensland. *Austral. J. Exper. Agric. Anim. Husb.* 7: 57–65.

JONES, R. J. & A. J. PRITCHARD, – 1971 – The method of reproduction in Rhodes grass (*Chloris gayana*). *Trop. Agric. Trin.* 48: 301–309.

JONES, R. J., A. A. SEAWRIGHT & D. A. LITTLE, – 1970 – Oxalate poisoning in animals grazing the tropical grass *Setaria sphacelata*. *J. Austral. Inst. Agric. Sci.* 36: 41–43.

JONES, R. K., P. J. ROBINSON, K. P. HAYDOCK & R. G. MEGARRITY, – 1971 – Sulphur-nitrogen relationships in the tropical legume, *Stylosanthes humilis*. *Austral. J. Agric. Res.* 22: 885–894.

JONES, R. M., – 1969 – Mortality of some tropical grasses and legumes following frosting in the first winter after sowing. *Tropical Grasslands*: 3: 57–63.

KIKKAWA, J. & W. T. WILLIAMS, – 1971 – Altitudinal distribution of land birds in New Guinea. *Search* 2: 64–65.

LAZARIDES, M., M. J. T. NORMAN & R. A. PERRY, – 1965 – Wet-season development pattern of some native grasses at Katherine, N.T. CSIRO Division of Land Research and Regional Survey Technical Paper no. 26. 19 pp.

LLOYD, D. L., – 1970 – Makarikari grass studies, Toowoomba, Queensland, In 'Proceedings Eleventh International Grassland Congress,' pp. 230–235. University of Queensland Press, St. Lucia.

LLOYD, D. L., – 1971 – Growth and development of Makarikari grasses and three other subtropical species grown in the field at Toowoomba, Queensland. *Austral. J. Exp. Agric. Anim. Husb.* 11: 525–531.

LUDLOW, M. M., – 1969 – Dry matter production, leaf angle, silica content and digestibility of tropical pasture grasses. *J. Austral. Inst. Agric. Sci.* 35: 200–201.

LUDLOW, M. M. & G. L. WILSON, – 1971a – Photosynthesis of tropical pasture plants. 1. Illuminance, carbon dioxide concentration, leaf temperature and leaf-air vapour pressure difference. *Austral. J. Biol. Sci.* 24: 449–470.

LUDLOW, M. M. & G. L. WILSON, – 1971b – 2. Temperature and illuminance history. *Austral. J. Biol. Sci.* 24: 1065–1075.

MCWILLIAM, J. R., – 1969 – Introduction, evaluation and breeding of new pasture species. *J. Austral. Inst. Agric. Sci.* 35: 90–98.

't MANNETJE, L., – 1965 – The effect of photoperiod on flowering, growth habit and

dry matter production in four species of the genus *Stylosanthes*. *Austral. J. Agric. Res.* 16: 767–771.
'T MANNETJE, L., – 1969 – Rhizobium affinities and phenetic relationships within the genus *Stylosanthes*. *Austral. J. Bot.* 17: 553–564.
'T MANNETJE, L. & A. J. PRITCHARD, – 1968 – The effects of photoperiod and night temperature on flowering and growth in some African *Trifolium* species. *New Phytol.* 67: 257–263.
'T MANNETJE, L. & N. H. SHAW, – 1972 – Nitrogen fertilizer responses of a *Heteropogon contortus* and *Paspalum plicatulum* pasture in relation to rainfall in central coastal Queensland. *Austral. J. Exp. Agric. Anim. Husb.* 12: 28–35.
MILFORD, R., – 1967 – Nutritive values and chemical composition of seven tropical legumes and lucerne grown in subtropical south-eastern Queensland. *Austral. J. Exper. Agric. Anim. Husb.* 7: 540–545.
MILFORD, R. & K. P. HAYDOCK – 1965 – The nutritive value of protein in subtropical pasture species grown in south-east Queensland. *Austral. J. Exper. Agric. Anim. Husb.* 5: 13–17.
MILLER, H. P., – 1967 – The effect of pre-treatment of native pasture and sowing rate on establishment of Townsville lucerne on Tippera clay loam at Katherine, N.T. *Australian J. Exper. Agric. Anim. Husb.* 7: 514–517.
MILLER, H. P. & R. A. PERRY, – 1968 – Preliminary studies on the establishment of Townsville lucerne (*Stylosanthes humilis*) in uncleared native pasture at Katherine, N.T. *Austral. J. Exper. Agric. Anim. Husb.* 8: 25–32.
MINSON, D. J. & M. N. MCLEOD, – 1970 – The digestibility of temperate and tropical grasses. In 'Proceedings of Eleventh International Grassland Congress,' pp. 719–722. University of Queensland Press, St. Lucia.
MINSON, D. J. & R. MILFORD, – 1966 – The energy values and nutritive value indices of *Digitaria decumbens*, *Sorghum almum* and *Phaseolus atropurpureus*. *Austral. J. Agric. Res.* 17: 411–423.
MOORE, R. M. (ed.), – 1970 – Australian Grasslands. Australian National University Press, Canberra. 473 pp.
MORLEY, F. H. W., – 1962 – Pasture plant breeding and animal production. *J. Austral. Inst. Agri. Sci.* 28: 3–7.
NICHOLAS, D. B., – 1971 – Genotypic variation in growth and nodulation in *Glycine wightii*. *J. Austral. Inst. Agric. Sci.* 37: 69–70.
NICHOLAS, D. B. & K. P. HAYDOCK, – 1970 – Variation in growth and nodulation of *Glycine wightii* under controlled environment. *Austral. J. Agric. Res.* 22: 223–230.
NORMAN, M. J. T., – 1969 – The effect of burning and seasonal rainfall on native pasture at Katherine, N. T., *Austral. J. Exper. Agric. Anim. Husb.* 9: 295–298.
NORTHERN TERRITORY, AGRICULTURAL BRANCH – 1966 – *Stylosanthes humilis* H.B.K. (*S. sundaica*). In 'Proceedings of Ninth International Grassland Congres,' pp. 1719–1722. Departamento do Produção Animal da Secretaria da Agricultura do Estado de São Paulo.
PAIJMANS, K. & E. LÖFFLER, – 1972 – High-altitude forests and grasslands of Mount Albert Edward, New Guinea. *J. Trop. Geogr.* 34: 58–64.
PERRY, R. A., – 1960 – Pasture Lands of the Northern Territory, Australia. Land Research Series, no. 5. CSIRO, Melbourne. 55 pp.
PERRY, R. A., – 1970 – The effects on grass and browse production of various treatments on a mulga community in central Australia. In 'Proceedings of Eleventh International Grassland Congress,' pp. 63–66. University of Queensland Press, St. Lucia.
PLUCKNETT, D. L., – 1970 – Productivity of tropical pastures in Hawaii. In 'Proceedings of Eleventh International Grassland Congress,' pp. A38–49. University of Queensland Press, St. Lucia.

PRITCHARD, A. J., – 1962 – The cytology and reproduction of *Paspalum yaguaronense* (Henr.). *Austral. J. Agric. Res.* 13: 206–211.

PRITCHARD, A. J., – 1967 – Apomixis in *Brachiaria decumbens*, Stapf. *J. Austral. Inst. Agric. Sci.* 33: 264–265.

PRITCHARD, A. J., – 1970 – Meiosis and embryo sac development in *Urochloa mosambicensis* and three *Paspalum* species. *Austral. J. Agric. Res.* 21: 649–652.

PRITCHARD, A. J., – 1971 – The hybrid between *Pennisetum typhoides* and *P. purpureum* as a potential forage crop in south-eastern Queensland. *Tropical Grasslands* 5: 35–39.

PRITCHARD, A. J. & L. 't MANNETJE, – 1967 – The breeding systems and some interspecific relations of a number of African *Trifolium* spp. *Euphytica* 16: 324–329.

PRITCHARD, A. J. & J. G. WUTCH, – 1964 – Chromosome numbers in the genus *Glycine* L. *Nature* (*Lond.*) 202: 322.

RITSON, J. B., L. A. EDYE & P. J. ROBINSON, – 1971 – Botanical and chemical composition of a Townsville stylo-spear grass pasture in relation to conception rate of cows. *Austral. J. Agric. Res.* 22: 993–1007.

ROSS, M. A., – 1969 – An integrated approach to the ecology of arid Australia. *Proc. Ecol. Soc. Austral.* 4: 67–81.

ROTAR, P. P., – 1970 – Variation in agronomic characteristics of *Desmodium intortum* (Mill.) Urb. and a related species. In 'Proceedings of Eleventh International Grassland Congress,' pp. 296–299. University of Queensland Press, St. Lucia.

RUSSELL, J. S. & A. W. MOORE, – 1970 – Detection of homoclimates by numerical analysis with reference to the brigalow region (eastern Australia). *Agric. Meteorol.* 7: 455–479.

SANTHIRASEGARAM, K., J. E. COALDRAKE & M. H. M. SALIH, – 1966 – Yield of a mixed subtropical pasture in relation to frequency and height of cutting and leaf-area index. In 'Proceedings Tenth International Grassland Congress,' pp. 125–129.

SHAW, N. H., – 1961 – Increased beef production from Townsville lucerne (*Stylosanthes sundaica* Taub.) in the spear grass pastures of central coastal Queensland. *Austral. J. Exper. Agric. Anim. Husb.* 1: 73–80.

SHAW, N. H., T. W. ELICH, K. P. HAYDOCK & R. B. WAITE, – 1965. A comparison of seventeen introductions of *Paspalum* species and naturalized *P. dilatatum* under cutting at Samford, south-eastern Queensland. *Austral. J. Exper. Agric. Anim. Husb.* 5: 423–432.

SHAW, N.H., C. T. GATES & J. R. WILSON, – 1966 – Growth and chemical composition of Townsville lucerne (*Stylosanthes humilis*). 1. Dry matter yield and nitrogen content in response to superphosphate. *Austral. J. Exper. Agric. Anim. Husb.* 6: 150–156.

SHAW, N. H. & L. 't MANNETJE, – 1970 – Studies on a spear grass pasture in central coastal Queensland – the effect of fertilizer, stocking rate and oversowing with *Stylosanthes humilis* on beef production and botanical composition. *Tropical Grasslands* 4: 43–65.

SHAW, N. H. & M. J. T. NORMAN, – 1970 – Tropical and sub-tropical woodlands and grasslands. In 'Australian Grasslands,' ed. R. M. MOORE, pp. 111–122. Australian National University Press, Canberra.

TOTHILL, J. C., – 1968 – Soil temperatures and seed burial in relation to the performance of *Heteropogon contortus* and *Themeda australis* in burnt native woodland pastures in eastern Queensland. *Austral. J. Bot.* 17: 269–275.

TOTHILL, J. C., – 1971a – A review of fire in the management of native pasture with particular reference to north-eastern Australia. *Tropical Grasslands* 5: 1–10.

TOTHILL, J. C., – 1971b – Grazing, burning and fertilizing effects on the regrowth of some woody species in cleared open forest in south-east Queensland. *Tropical Grasslands* 5: 31–34.

WACE, N. M., – 1971 – Naturalised plants and native vegetation in Australia. Australian National University, Canberra. 2 pp.

WHITEMAN, R. C., – 1968 – The effects of temperature on the vegetative growth of six tropical legume species. *Austral. J. Exper. Agric. Anim. Husb.* 8: 528–532.

WHITEMAN, P. C., – 1970a – Seasonal changes in growth and nodulation of perennial tropical pasture legumes in the field. 2. Effects of controlled defoliation levels on nodulation of *Desmodium intortum* and *Phaseolus atropurpureus. Austral. J. Agric. Res.* 21: 207–214.

WHITEMAN, P. C., – 1970b – 3. Effects of flowering on nodulation of three *Desmodium* species. *Austral. J. Agric. Res.* 21: 215–222.

WHITEMAN, P. C. & P. GILLARD, – 1971 – Species of *Urochloa* as pasture plants. *Herbage Abstracts* 41: 351–357 (65 references).

WHITEMAN, P. C. & A. LULHAM – 1970 – Seasonal changes in growth and nodulation of perennial tropical pasture legumes in the field. 1. The influence of planting date and grazing and cutting on *Desmodium uncinatum* and *Phaseolus atropurpureus. Austral. J. Agric. Res.* 21: 195–206.

WILLIAMS, C. H. & C. S. ANDREW, – 1970 – Mineral nutrition of pastures. In 'Australian Grasslands,' ed. R. M. MOORE, pp. 321–338. Australian National University Press, Canberra.

WILSON, B., – 1968 – Pasture Improvement in Australia. K. G. Murray, Sydney, Melbourne, London. 288 pp.

WINKWORTH, R. E., – 1963 – The germination of buffel grass (*Cenchrus ciliaris*) seed after burial in a central Australian soil. *Austral. J. Exper. Agric. Anim. Husb.* 3: 326–328.

WINKWORTH, R. E., – 1969 – Germination of Townsville lucerne (*Stylosanthes humilis* H.B.K.) in relation to weather at Katherine, N.T. *J. Austral. Inst. Agric. Sci.* 35: 201–204.

WOODS, L. E., – 1969 – A survey of Townsville stylo (Townsville lucerne) pastures established in the Northern Territory up to 1969. *Tropical Grasslands* 3: 91–98.

WOODS, L. E. & R. A. DANCE, – 1970 – Seed and nutritional aspects of grass-Townsville stylo competition. *J. Austral. Inst. Agric. Sci.* 36: 45–47.

INDEX: SUBJECTS

INDEX: GEOGRAPHICAL NAMES

INDEX: PLANT NAMES

(Nomenclature as used by authors)